IN SERVICE TO AMERICAN PHARMACY

History of American Science and Technology Series

General Editor, LESTER D. STEPHENS

IN SERVICE TO AMERICAN PHARMACY

The Professional Life of William Procter, Jr.

GREGORY J. HIGBY

The University of Alabama Press

Tuscaloosa and London

Library of Congress Cataloging-in-Publication Data

Higby, Gregory.
    In service to American pharmacy : the professional life of William
Procter, Jr. / Gregory J. Higby.
    p.   cm. — (History of American science and technology series)
    Includes bibliographical references and index.
    ISBN 0-8173-0591-2 (alk. paper)
    1. Procter, William, 1817–1874.   2. Pharmacists—United States—
Biography.   I. Title.   II. Series.
    [DNLM: 1. Procter, William, 1817–1874.   2. Pharmacists—biography.
3. Pharmacy—history—United States.   WZ 100 P9642H]
RC73.P76H54   1992
615'.1'092—dc20
[B]
DNLM/DLC
for Library of Congress   92-406

        British Library Cataloguing-in-Publication Data available

Succeed in your business—be thorough masters of it—and then, if your taste inclines, and your talents fit you for scientific pursuits follow them . . . not for mere fame . . . but let the motive be the intrinsic pleasure which you derive from them, and the usefulness your labors contribute to abstract science, and the practical interests of man.

WILLIAM PROCTER, JR.,
Valedictory Address to the Graduates
of the Philadelphia College of Pharmacy,
Delivered in Sansom Street Hall, March 18, 1852

Contents

Illustrations

This engraving of Procter, based on an unpublished photograph, was the most popular portrait among his old friends and colleagues. It shows him in his prime during the 1850s and early 1860s. (From Kremers Reference Files, F. B. Power Pharmaceutical Library, University of Wisconsin-Madison School of Pharmacy, Madison, Wisc.)

Preface

The present position of the pharmacist in the structure of health care in the United States evolved during the middle half of the nineteenth century, roughly from the founding of the Philadelphia College of Pharmacy in 1821 to the passage of meaningful pharmaceutical legislation in the 1870s. During that period a small group of men connected with the drug trade championed the concept of a practitioner whose raison d'être was the preparation of medicines. This book describes and evaluates the progress of that movement through the contributions of William Procter, Jr. (1817–74), its predominant figure.

When Procter entered service as an apprentice apothecary in the early 1830s, the idea that a man could make a living largely by filling prescriptions was a new one in the United States—although centuries old in Europe—and it then was realized in practice only in a few urban centers. Little more than a decade had passed since some apothecaries and wholesale druggists in Philadelphia had organized the first pharmaceutical organization in the United States, the Philadelphia College of Pharmacy. As in the rest of the country, pharmacy in Philadelphia was still in a primitive state. Almost all who sold medicines did so as a trade that could be exchanged for another if more profitable. To them, selling drugs seemed no different from dealing in bread or candles, the object being to "buy cheap and sell dear." Merchants sold highly potent or poisonous drugs as casually as tea or salt.[1]

By the 1870s the practice of pharmacy in the United States had progressed to the point where drugs and medicines were handled in most cities and towns by men who had received some pharmaceutical training. Formal instruction in pharmacy was available in several urban areas. The quality of the products sold in drugstores had improved, and physicians were coming to rely upon pharmacists as dispensers of prescription medicines. States and localities had begun to pass laws that required the registration of practitioners and in some cases examined their competency. With their vocational identity established, a new stratum of pharmacists, largely trained and educated in the pioneer schools of pharmacy, started a new movement in the late nineteenth century that sought to elevate the reputation of their calling in American society to a level approaching that of established professions.

William Procter, Jr., more than any other figure provided leadership during this developmental period in the history of American pharmacy through a variety of professional and scientific activities, and it is this side of his life that I examine in this book. After an introduction that outlines the status of pharmacy before Procter and a chapter exploring his character, I turn to six distinct areas of his career that relate directly to the development of American pharmacy. Chapters 2 through 7 are in a sense self-contained studies of pharmaceutical practice, science, quality control, journalism, education, and organizations during the middle half of the nineteenth century. The topical arrangement is intended to help clarify and evaluate Procter's contribution to pharmacy, and is ordered to correspond roughly to the sequence of his career. I begin with Procter's drugstore, his base of operations. From there he conducted pharmaceutical investigations and drug assays, edited the *American Journal of Pharmacy,* composed lectures, and performed his duties as corresponding secretary of the young American Pharmaceutical Association. Some redundancy usually accompanies a purely topical treatment of a person's life and this biography is no exception. Yet, by repeating information when needed, each chapter provides a clear description of Procter's views and activities for those interested in particular aspects of American pharmacy in the nineteenth century.

I came to study Procter in a roundabout way. My main research interest has centered on the period between 1875 and 1920, when American pharmacists developed a confident self-identity within a rapidly changing health-care marketplace. It was an era of camaraderie and stiff competition, of scientific expeditions and soda fountains. State laws prevented the grossly incompetent from practicing, but did not impose restrictive educational requirements. Some of the most divisive controversies concerning pharmaceutical practice arose or reached a crisis during this time: drug substitution, pharmacist prescribing and doctor dispensing, price cutting, chain drugstores, and the entry of women into the profession.

To do historical work in this time frame, I needed to understand how and why American pharmacists found themselves in this complex situation in the years around 1900. In my background research on the early period of American pharmacy's professionalization (1820–75), I came across Procter's name and writings again and again. His was the voice of this period. I was surprised to find that all biographical treatments of this key figure were based on two limited sources: a twenty-one-page memoir composed by Charles Bullock in 1874 and a few brief reminiscences written by Joseph Remington around 1900.[2] Bullock wrote about a close friend who had died suddenly, and so was not critical of Procter. Remington's "pen portraits" of his former teacher are an entertaining mixture of hero worship and nostalgia. Together they provide details about how Procter practiced and taught pharmacy, as well as about the character of his in-shop laboratory, that are unavailable elsewhere. Moreover, these sketches demonstrate the large influence Procter had on Remington, the predominant figure in American pharmacy from 1880 to 1920.

Yet these sources put forth an incomplete account of Procter's achievements. They fail to delve into the details of his diverse career and to evaluate his influence on future professional and scientific developments. A series of chats with Glenn Sonnedecker convinced me that I should tackle the Procter biography myself. The result of that labor is this book.

You may wonder why, if Procter is so important, this is the first book-length biography of the so-called father of American pharmacy. No doubt the almost total lack of surviving personal papers has discouraged others from attempting the task.[3] Still, upon close examination I found a voluminous printed record in the *American Journal of Pharmacy*—which Procter edited during his prime—and in the *Proceedings* of the American Pharmaceutical Association, as well as in a few previously untapped manuscript sources. The latter were turned up through a circular letter sent out to several historical societies, manuscript repositories, and other similar institutions. Collections at the Philadelphia College of Pharmacy and Science, the College of Physicians of Philadelphia, and the American Pharmaceutical Association headquartered in Washington, D.C., were most valuable. Ironically, after searching far and wide, the best cache of Procter manuscripts was found in Kremers Reference Files at the University of Wisconsin School of Pharmacy in Madison. This cache consists of sixteen letters from Procter to Albert Ebert of Chicago. They came into the hands of Edward Kremers around 1900 when the American Pharmaceutical Association's Section on Historical Pharmacy put out a call for all letters written by Procter and other prominent figures of the association.

In the absence of sufficient personal material, I decided to write a "professional biography," that is, a treatment of Procter's career accomplishments. And yet, after long hours of reading his editorials and papers, a special and unique voice emerged from the printed pages. It is my hope that some of his character and drive comes across in the pages that follow.

One caveat is in order: When reading this book, it is imperative to see Procter as part of an urban elite that sought to raise American pharmacy from a common trade to a profession respected by physicians and the public alike. Procter and his colleagues in this movement emulated the position, prestige, and scientific prowess exhibited by some continental pharmacists of the late eighteenth and early nineteenth centuries. Through their leadership, first in local and later in national organizations, this small group of men composed much of the educational program and legislation that molded later generations of pharmaceutical practitioners in the United States. The position of the American pharmacist as the scientific guarantor of drug quality and expert preparer of medicines—the dominant missions of the calling until the last third of the twentieth century—was formulated by these pharmaceutists and druggists of the mid-nineteenth century. Procter's contribution deserves examination because of its salient role in this effort; because of its exceptional nature, however, Procter's career should not be interpreted as indicative of common pharmaceutical practice in America during the mid-nineteenth century.

Acknowledgments

I would like to acknowledge a few people who have helped me a great deal. I wish first of all to thank Glenn Sonnedecker for his kind assistance and friendly encouragement throughout this project. I also thank John Parascandola and Ronald Numbers for their comments on an earlier version of this manuscript. (Someday, as my research and writing matures, I may answer most of their concerns.) The critical remarks of James Harvey Young and an anonymous reader were of great help. I am grateful for the financial support received from the University of Wisconsin's School of Pharmacy and the American Institute of the History of Pharmacy. In addition, this effort would have been almost impossible without the excellent collections in the history of pharmacy located and maintained at the University of Wisconsin. The F. B. Power Pharmaceutical Library and its staff supervised by Delores Nemec deserve special mention. I thank the fine staff of the Philadelphia College of Pharmacy and Science, both in the dean's office and in the England Library. Special thanks go to Nancy Weinstock, their former rare books librarian. In addition, I would like to thank the archivists and historians at the Smithsonian Institution for their courteous aid, above all Marc Rothenberg, who saved me hours of searching. The warm reception from George

Griffenhagen at the American Pharmaceutical Association made my work there a pleasure and solidified my professional commitment to that organization. I must acknowledge the support of several old friends: John Swann, who listened to my scholarly problems; Jeffrey Van Fleet, who listened to my early personal frustrations; Frank Cuva, who listened to me at the end; and Greg Melton, who helped with some of the typing. As always, Elisabeth Lambright has been a source of inspiration. Rosemary Zurlo-Cuva and Elaine Condouris Stroud pushed me hard to finish and deserve, therefore, some of the blame. Last, I offer my deepest gratitude to my dear wife, Marian Fredal, who never let me down.

IN SERVICE TO AMERICAN PHARMACY

Introduction

As a profession pharmacy does not excite or inspire. Instead, the word *pharmacist* elicits an image of a white-jacketed shopkeeper behind the counter of a small- to medium-sized store. Pharmacy, especially in the United States, has never produced hero-practitioners like Walter Reed or the fictional Martin Arrowsmith. American pharmacy possesses no scientist-saviors such as Jonas Salk or national authorities like Benjamin Spock. Like other secondary professions, pharmacy has its own quiet heroes and heroines, men and women who have contributed without widespread fame and usually without an accumulation of wealth.

The subject of this book, William Procter, Jr., made no fortune, started no great pharmaceutical firm, discovered no new wonder drug, founded no school or journal. Yet, upon his death in 1874, he was hailed as the father of American pharmacy, a title never challenged since. He earned that distinction from his wholehearted and complete dedication to what he called the "progress of American pharmacy." A quiet, private man, Procter burned with an ambition to raise his chosen calling, pharmacy, from its position as a disorganized trade to professional status. In pursuit of this goal Procter spread his vocational activities over almost all aspects of pharmacy. His life was his work, and so it is that each major area of his work is treated in a separate chapter in this book.

I

I chose Procter because his exemplary career illustrates well the diverse elements in flux that made up American pharmacy in the mid-nineteenth century. Moreover, his efforts to raise pharmacy to a professional level pinpoint several of the problems that beset the young occupation. Procter's lifetime (1817–74) coincides almost perfectly with American pharmacy's first phase of professionalization (c. 1820–75), that is, from the founding of the first pharmaceutical associations to the enactment of the first effective state laws regulating pharmacy. A large part of what follows evaluates how much Procter influenced his occupational milieu and vice versa.

Among the other leaders of the young occupation of pharmacy, Procter stood out. There may have been better businessmen, more inspiring teachers, or more popular authors, but he alone had nearly universal respect. Why? Because he kept all of his various efforts focused on his goal of professional improvement. Moreover, his modest approach, leading by the pen and by example, stood out in a time marked by big talk and brash actions.

Procter's concept of professionalism was not exceptional for mid-nineteenth-century America: that each individual practitioner gains professional stature in his community through the high quality of his occupational practice and his personal morality. The major purpose of group interaction and organizations was to improve individual practice via the sharing of new scientific and technical advances. As his authority grew, Procter added his voice to the vast majority of pharmacists who clung to this approach to professionalization in the face of educational and occupational reform that occurred following the Civil War. Thus he deserves much of the credit (and the blame) for what American pharmacy became in the late nineteenth and early twentieth centuries: an occupation that had gained public respect while split internally by differing approaches to professionalization.[1]

Professionalism and American Pharmacy

Because *professional* and *professionalism* were at the center of Procter's efforts, these terms are used with care. Throughout this book,

I avoid generic definitions of professionalism, which are usually based on the traditional learned professions of medicine, divinity, and law.[2] Eliot Friedson has argued effectively that the crux of professionalization studies "lies in embracing the concept [of professionalism] as an intrinsically ambiguous, multifaceted folk concept, of which no single definition and no attempt at isolating its essence will ever be generally persuasive. . . . [T]here is no single, truly explanatory trait or characteristic . . . that can join together all occupations called professions beyond the actual fact of coming to be called professions."[3]

Instead I follow the advice of sociologist Everett C. Hughes, who wrote that "the way to understand what professions mean in our society is to note the ways in which occupations try to change themselves or their image, or both, in the course of a movement to become 'professionalized.'"[4] In other words, I have tried to avoid the pitfall of applying the attributes of today's professionalism to practitioners of the last century. That common error, according to Donald M. Scott, "tends to reify mastery of a clearly delimited cognitive area and the existence of an autonomous community of the competent as the basic characteristics of a profession; it also tends to impose on the past a definition of profession that often bears little resemblance to many past professional groups." Instead, Scott argues that for historians "it is more useful to approach 'profession' as a changing social and cultural construct, encompassing different meanings of the idea of profession and different institutional arrangements at different times."[5] Rather than projecting twentieth-century professional values back on nineteenth-century practitioners of pharmacy, I have sought out their own views on professionalism, focusing on the most important protagonist of such ideals during this critical transitional phase, William Procter, Jr.[6]

Before turning to an analysis of Procter's attitudes toward the progress of American pharmacy during the middle third of the nineteenth century, it is necessary to set the scene. What was the state of pharmacy in the United States in the early nineteenth century?

Emergence of American Pharmacy

Between the War of 1812 and the Civil War a new occupation carved out its niche in American life: what we today call pharmacy. Although pharmacy had been a recognized profession in the Western world for centuries, it did not gain a place in American society until the years just before the Civil War. It arose out of the complex structure of the Atlantic Coast drug trade, which consisted of drug importers, wholesalers, manufacturing chemists, and apothecaries.[7]

Cast of Characters

In the United States no single profession or occupation has ever controlled the distribution of drugs and medicines. Just as today, shopkeepers in colonial times sold tonics and cough medicine. In the early nineteenth century, a wide variety of workers practiced *pharmacy,* that is, the compounding of medicines from crude drugs or ingredients.[8] In order to make the following discussion fairly clear, I will introduce the cast of characters involved in pharmaceutical practice, before concentrating on changes that occurred in pharmacy.

In the antebellum period, American medical care was in a state of disarray, with a plethora of poorly trained physicians (both regular and irregular), itinerant doctors, traveling peddlers of drugs, and proprietors of drugstores dispensing different levels of advice over the counter. Moreover, domestic medicine treated the vast majority of ailments in a nation predominantly rural and generally skeptical of the value of physicians. This was especially true before widespread industrialization undermined the self-reliance of most Americans. Aside from the sizable minority who advocated drugless treatments such as dietary reform and hydrotherapy, most purveyors of health care relied on some form of medicine. Because of this, drugstores, either wholesale or retail, served as a point of intersection for most Americans concerned with healing. The coincidental growth in the number of drugs available during the antebellum period also helped to stimulate the development of the American drug market.[9]

Most of the botanical drugs used in the United States in the nineteenth century were not cultivated in significant quantities in North America and had to be imported: asafetida from Iran, camphor from the East Indies, opium from Turkey, and cinchona bark from South America. Most drugs came to the United States through general merchants rather than through specialty houses before the Civil War. Brokers would then sell the imported drugs to wholesalers for a small commission. Only a few drug houses, like Schieffelin Brothers of New York, employed foreign buyers to ensure the quality of imported drugs. Without this check, most drug wholesalers were forced to fight a continuing battle of wits with clever brokers trying to palm off poor quality or adulterated drugs. This concern for quality stimulated many of the scientific and technological investigations of the antebellum period as well.

Drug Wholesalers

For those not familiar with the history of American pharmacy, the term *druggist* can be confusing. Within the drug trade itself, the term has generally applied to wholesale distributors of drugs. Yet these same businessmen often sold retail as well, confusing matters. Moreover, in the American vernacular, retailers of drugs and medicines have been called druggists. For clarity's sake in the next few paragraphs, I will use instead another common term of the nineteenth century for such middlemen: *jobber.* Throughout this book, however, *druggist* will identify a wholesaler of drugs.

The drug jobber sometimes imported his own drugs and medicines, but usually bought them from an importer or broker. His clientele can be divided into two broad categories: those who bought in person and those who communicated by mail or telegraph. General wholesalers from the interior traveled once or twice a year to Boston, Philadelphia, or New York to place large orders, usually spreading out their business among several jobbers. In addition, local retail apothecaries came in person to place stock orders and look for bargains. By visiting in person they could judge the quality of the merchandise before buying. In contrast, country physicians and storekeepers (including retail drug sellers) usually corresponded with jobbers. Because they could not examine medicines before purchase, small buyers from the interior usually dealt with only one jobber and on credit of six months. Jobbers guarded their reputations closely since they relied heavily on the trust of physicians and retailers in the West. By the late 1850s, the drug trade developed to the point where jobbers specialized by serving one or two segments of the drug distribution network. [10]

Out where drugstores and doctors were far between, country general stores sold a wide variety of drugs and medicines. Advertisements in newspapers placed by big city jobbers promised the highest quality cinchona bark, opium, jalap, calomel, ipecac, and chamomile. Of course, patent medicine orders were "thankfully received." In an economy starved for currency, bartering dominated country store business. Consequently, jobbers back in Philadelphia or New York had to be ready to accept farm produce in exchange for their merchandise. [11]

Physicians, especially those who did not keep shops but just ran an office or part-time practice, were in a more difficult position. They rarely traveled to the big cities to order drugs from jobbers and relied on cash transactions or, at best, short-term credit. The doctoring business, however, was notoriously cash poor, with almost all services done on credit. It is not too surprising then that physicians in the interior generally welcomed the arrival of drugstores to their localities. When a physician wrote out his prescriptions, his fees were for service only and he could spend less of his hard-to-obtain currency on drug supplies.

For physicians and domestic healers, country apothecaries served as secondary wholesalers of drugs and medicines as well as general retailers. Like other specialized retailers, apothecaries required a critical mass of population and currency, and therefore arrived in a settlement only after it was well established.

In the country, the "dividing line between a drugstore and a general store was often not distinct. Drugstores sold dye-woods, turpentine, alcohol, liquor, paints and varnishes, glassware, often a line of notions and some groceries, non-proprietary drugs such as calomel and quinine—and of course the patent medicines. So did the general store."[12] And the grocery line of general stores carried spices and herbs like cloves and mints used in domestic healing. Patent medicines were especially attractive to general store keepers because of their high markup, or what was known in the retail trade as "regular drug store profit."[13]

Apothecaries were not especially fond of competition from general stores and tended to couch their criticisms in sarcastic language. In an editorial Procter wrote, "In our country villages . . . a large amount of medicines are sold by country store keepers who know as much about [cinchona] bark, rhubarb, and opium, as they do about algebra and conic sections."[14]

Physicians

In the antebellum period, chaos came to the profession of medicine. The traditional practice of medicine and its education and regulatory systems were challenged by groups of alternative healers, usually referred to as "sectarians" by historians. Some sectarian groups had origins in Europe (homeopaths) or others advocated drugless therapies (hydropaths), but most were home-grown healers convinced of the superiority of herbal treatments over the mineral-laden materia medica of regular physicians. In 1854 Procter reflected on the history of this "American botanico-medical movement," which influenced the direction of pharmacy:

About forty years ago, Samuel Thomson, of New England, an energetic but illiterate man, commenced the practice of that system of empyricism that subsequently under the name of Thomsonianism was seized upon by the popular mind, and for a time became the favorite practice of a numerous class of persons, especially among farmers, who, pleased with the idea of being their own physicians, were not slow in patronizing a scheme that, without collegiate study, offer[ed] to give them the knowledge requisite for medical practice. In the process of time the crude ideas of the founder were more and more modified by his disciples . . . and some degree of science, both as regards botany and pharmacy, crept into their practice, which now included a numerous list of the plants indigenous to our country. The advent from time to time of a regular physician among them brought in an admixture of regular ideas; and at this time, quite a numerous body of men, principally in the West and North, are engaged in the practice of a scheme of medicine and pharmacy which is known as "Eclecticism," or "the Eclectic Practice of Medicine."[15]

In order to compete with regular physicians, sectarians banded together and convinced state legislatures to repeal licensing laws. In Jacksonian America, the idea that "every man could be his own doctor" carried the day. Anyone with enough drive and courage could practice as a physician. In eastern urban centers, bastions for orthodox medical practice and education, sectarians were in the minority. The enthusiasm for indigenous drugs and overblown claims of achievements in scientific pharmacy common among botanical sectarians, however, did influence the elite of pharmacy in the East. Protective of their intellectual domain, they tested each new "discovery" of the botanics, particularly those dealing with indigenous drugs. They found most wanting, but in contrast to the elite of regular medicine, these pharmacists did not reject the efforts of botanics out of hand.[16]

Until the late nineteenth century, American physicians, whether sectarian or regular, were forced by circumstances and convention to compound and dispense many of their own prescriptions. Pharmacists were common in cities and towns, but not common in areas of sparse population. At the beginning of the nineteenth century, a medical apprentice's first introduction to physick usually came in the compounding of his preceptor's prescriptions. The acquisition of manual skills, such as the pharmaceutical and surgical arts, was a prime objective of apprenticeship.[17] As medical school education displaced the apprenticeship system during the 1820s and 1830s, the pharmaceutical knowledge of physicians began to decline, with physicians relying more and more on apothecaries and later on drug manufacturers.

Because most antebellum physicians practiced far from any drugstore, the filling of medicine bags and chests was a major source of business in the drug trade. Some jobbers and large retailers like the Parrish brothers of Philadelphia specialized in their production.[18] In the nineteenth century the offices of physicians contained a wide gamut of pharmaceutical furnishings, running from a shelf or two of standard preparations to fully equipped shops. Early in the century, when the apprenticeship system still ruled, the "pharmacy" was manned by a tyro physician. Coming out of the tradition of the British apothecary shop, so-called doctor's shops commonly provided pharmaceutical services in the late eighteenth and early nineteenth centuries. Dr. John Morgan, the best-known advocate of separating pharmacy and medicine in the young republic, eventually returned to the shop himself to make ends meet.[19]

As pharmacies became more common as part of a general trend toward specialty retail establishments, doctor's shops declined in the first quarter of the nineteenth century. Yet, as young physicians poured out of proprietary medical schools in the second quarter of the century to find urban areas full of competitors, many resorted to opening shops. Because they appeared to charge only for the medicine they prescribed and not for their services, these doctors were cursed by established physicians and apothecaries alike.

Shop doctors tended to hire employee apothecaries (usually called "drug clerks") for two reasons. If they had missed out on much of the pharmaceutical training usually obtained through a long medical apprenticeship, young physicians needed someone to compound prescriptions. Even if they had sufficient expertise, successful shop doctors sometimes hired a drug clerk to keep shop while they diagnosed and prescribed. If they succeeded to the point of having a full office-style practice, shop doctors often sold off their businesses to their clerks, thereby stimulating the growth of the retail drug trade. As proprietors, former drug clerks took on the title of apothecary.

Although former clerks at doctor's shops added to the number of practicing apothecaries, most had roots in drug wholesaling. In the backrooms and cellars of drug warehouses young men learned pharmacy—"the art . . . of preparing, preserving, and compounding substances . . . for the purposes of medicine."[20] Generally speaking, there were relatively few specialized retail establishments in the United States before 1815. Drugstores commonly combined a small retail business with a more sizable wholesale trade, but with separate departments: "the front part of the lower floor devoted to retail and prescription business, and the rear to the counting-room and packing of goods for city and country trade." The upper floor or floors served as warehouses, with chemical work done in the basement.[21]

Apothecaries

Apothecaries had practiced in North America since the early colonial days, but American physicians in general did not view their services as essential and distinct until the early nineteenth century. Physicians or their apprentices compounded almost all prescriptions with drugs purchased from wholesalers or from the small number of retail drug sellers (apothecaries) who practiced in urban areas. A look at the policies of early American hospitals illustrates a shift in attitudes. The first hospitals of the young republic, for instance, employed medical apprentices as staff apothecaries. *A Brief Account of the New-York Hospital,* published in 1804, for instance, states: "A house Surgeon and Apothecary constantly reside in the Hospital.—these offices are filled by the students of the Physicians and Surgeons belonging to the Hospital, which affords an excellent school for the young men appointed to those places."[22] The staff apothecary practiced both pharmacy and medicine in a manner analogous to the British apothecary, going on rounds, treating patients, and compounding prescriptions.

By 1811, however, the position of apothecary at the hospital had changed. The person chosen was a full-time pharmaceutical practitioner, tested before hiring on his prowess as a compounder of medicines. Instead of being obligated to go on rounds, he was required to stay in his "shop" (workroom) at all times.[23] By 1819 the services of the pharmaceutical apothecary had become so essential that the hospital's bylaws not only required testimonials on an applicant's behalf, but also a $250 bond to ensure "faithful performance of the duties of his office, and that he will not cease to perform the duties of this office, without giving two months notice of his intention to leave his employment."[24]

Dispensaries followed a similar pattern, replacing medical apprentices with full-time apothecaries. By selecting responsible, mature individuals as apothecaries, the directors of hospitals and dispensaries could obtain competent day-to-day management of their institutions. (The apothecary usually served in multiple capacities, such as managing accounts and running the library.) More importantly for the budding occupation of pharmacy, full-time apothecaries provided reliable pharmaceutical services and probably encouraged prescription writing. From the 1820s on, as more and more physicians gained their clinical experience in hospitals and dispensaries instead of with preceptors, they learned to write prescriptions, rather than to compound them.[25] After graduation some young physicians continued to write out prescriptions, thereby stimulating the growth of pharmacy.

In the years before and after the War of 1812, urban physicians came to depend more and more on the expertise of apothecaries to ensure the quality of drugs and medicines. Yet even the most common of preparations such as laudanum or tincture of digitalis were made in a wide variety of ways, following recipes culled from European reference books.[26] The appearance of American books of drug standards, such as the *Pharmacopoeia of the Massachusetts Medical Society* in 1808 and the *Pharmacopoeia of the United States of America (USP)* in 1820, reflected both this concern and the increasing complexity of the drug market. Prescribing physicians organized their efforts to produce these guidebooks for apothecaries to ensure uniformity in the preparation of medicines. The preface to the *Massachusetts Pharmacopoeia* refers to "a perfect understanding [that] should exist."[27]

After the War of 1812 urban physicians continued to dispense, but people began to take advantage of the slowly growing number of retail apothecaries. For example, the fee bill approved by the New-York County Medical Society in January 1816 contained a detailed section of "Pharmaceutical Charges," while in Boston, which had a more developed cadre of apothecaries, the medical association omitted pharmaceutical charges from its fee bill.[28] The number of doctor's shops, reflected in advertisements and city directories, was dropping off rapidly. A small class of retail apothecaries presented no particular threat to urban physicians in the first decades of the nineteenth century and provided several conveniences.

When examining the variety of practitioners involved in the production and sale of medicines before the Civil War, one should remember how easily occupations were exchanged as circumstances warranted. For example, around 1805 an Englishman named Constantine Adamson entered an apothecary's shop in London as an apprentice at the rather late age of 20. After his apprenticeship he went to Nova Scotia to make his fortune. Adamson could not get a business going and subsequently joined the provincial army in 1812. After the war with the United States, he entered the Canadian lumber trade, where he made and lost a fortune in wood. Apparently to escape creditors, Adamson traveled to New York in 1817 and began teaching school. His health failed and he was treated by the physician-druggist Richard Seaman of the firm Walters and Seaman. Seaman's business failed and he convinced Adamson to be his partner in a new store. Adamson returned to the drug trade after ten years and built a reputation as one of New York's most honest and scientific druggists. He died in 1846 at the age of 62. His career as apothecary, soldier, lumber baron, schoolteacher, and druggist was not viewed as unusual.[29]

Situation in 1820s and 1830s

East Coast apothecary shops became more standard in their appearances and in the stock they carried during the 1820s and 1830s. The observation of Porter and Livesay that specialization among merchants in the early nineteenth century "resulted from conservatism, not innovation" applies well to the drug trade. Merchants, jobbers, and retailers of drugs and medicines dropped parts of their general lines to concentrate on those that seemed to promise the most expeditious profit. As part of the rapid rise of specialty shops, apothecary shops in the 1820s and 1830s tended to limit their merchandise a bit more to drugs, medicines, surgical supplies, artificial teeth and limbs, dyestuffs, essences, and chemicals. Exotic dietary items such as figs, raisins, and citrus fruits were taken over by grocers.[30] Drugstores in small cities and in towns, however, tended to keep in stock more general articles such as glass, paints, varnishes, and oils.[31]

The educated elite of Atlantic Coast physicians fostered the development of a well-trained, yet subservient pharmaceutical profession in the 1820s and 1830s. They welcomed the early pharmaceutical associations and served as faculty for the first pharmacy schools. Lectures given by these physician–educators reveal an encouraging, paternalistic attitude probably held by most physicians toward the young pharmaceutical profession.[32]

In the introductory editorial of the new *Journal of the Philadelphia College of Pharmacy,* the publishing committee of four physicians and one pharmacist wrote: "The history of the progress of society is that of the division of labour, and there is no surer indication of advancement in the arts of civilization, than the multiplicity and subdivision of occupations." They contended that this progress had led to the separation of the special work of the apothecary from that of the "drug factor, the druggist, the manufacturing chemist, the drug powderer, the paint and oil dealer, [and] the varnish maker."[33] Physicians voiced support for the growth of an independent profession of pharmacy as a "necessity for a division of labor" to meet the "growing demands of the community."[34]

Some druggists and apothecaries demonstrated their expertise in the new pharmaceutical arts and sciences, thereby earning respect among physicians for the young pharmaceutical calling. In the late 1820s some of these men took the title "pharmaceutist" in an apparent effort to separate themselves from the older generation of apothecaries and druggists. Oddly enough, the term seems to have caught on better among educated physicians of the 1830s and 1840s than it did among pharmaceutical practitioners themselves.[35]

Besides their skills making preparations, apothecaries demonstrated their worth to physicians by acting as an additional check for drug quality. As Europeans became more concerned with drug quality, they used America as a dumping ground for weak or adulterated drugs. In addition, alkaloidal chemistry allowed European firms to extract the active constituents from botanicals and export the exhausted remains to foreign markets. Pharmaceutists who stood behind the potency of their medicines gained patronage among urban physicians.[36]

As alluded to previously, the number of physicians trained in medical schools rather than by preceptors grew rapidly in the 1820s and 1830s. Medical students attended lectures and read textbooks that strongly emphasized therapeutics, but lacked significant pharmaceutical content. Graduates who went to the interior depended heavily on medicine chests prepared by jobbers, while those who stayed in cities wrote more prescriptions and dispensed fewer medicines themselves.

The general growth and specialization of the economy in the 1820s and 1830s also helped apothecaries make it in business. Apothecary shops became the main distributors of patent medicines, a highly profitable line of merchandise. Even circumstances disastrous to other businesses, such as the cholera epidemic of 1832, helped drugstores.[37]

The relationship between urban doctors and druggists began to sour in the 1840s. A physician in the mid-1830s predicted future trends well when he wrote: "The apothecaries of Boston, without exception, are an intelligent, high-minded, and honorable class of men, to whom the community looks with confidence; but population is rapidly increasing, apothecaries will multiply, and unless some spirited method is speedily devised for teaching the successive generations which are constantly graduated from behind their master's counters, the country will be overrun with a shoal of ignorant, presuming apothecaries, who will prosper just in proportion to the misery they create."[38]

Before the 1840s physicians rarely made negative comments concerning the practices of apothecaries or druggists. Even in the diatribes against nostrums so common in medical journals, physicians did not blame apothecaries for the popularity of these "quack medicines." In cases of poisoning, also, physicians did not blame apothecaries for indiscriminate sales practices.[39]

Clashes between physicians and apothecaries did not occur commonly before the 1840s largely because apothecaries accepted their subservient position to physicians, and the number of drugstores, although on the rise, had not yet reached the saturation point. Feeling more confident of their standing, apothecaries began shifting their allegiance from physicians to their customers.

Consequently, they took to refilling prescriptions without physician authorization, and to diagnosing and treating customers, a practice called "counter prescribing."

In large cities doctor's shops were back on the rise after a decline of three decades. Medical schools continued to turn out hundreds of graduates, most of whom sought their fortunes in urban areas. Finding the going difficult, they would sometimes turn to retailing their skills.[40] As was the case earlier in the century, successful shop doctors hired drug clerks to keep shop as they went on their rounds. These clerks would often buy into the business and become apothecaries, adding to the rate of growth in the retail drug trade.

William Procter, Jr., entered into this professional chaos when he opened his own shop in Philadelphia in 1844. He had begun his apprenticeship in the early 1830s, when the position of urban pharmacy was much more secure. Philadelphia apothecaries like Elias Durand were regarded as full members of the local scientific community. European pharmacists were on the forefront of discoveries in chemistry. Pharmacy organizations in Philadelphia, New York, and Boston interacted with medical societies as near equals. But by the late 1830s Procter watched pharmacy's collective standing begin to deteriorate after seeing how close it had come to gaining widespread stature as a profession. As his small business faltered and nearly failed, Procter spent his free time indulging his scientific curiosity or considering the fate of his chosen calling. During these years of quiet struggle, he formulated a program for progress based on education and scientific investigation with which to elevate American pharmacy. Before looking at how he tried to meet this challenge, in chapter 1 the text turns to the largely unknown personal side of Procter, which reveals to a small extent the drive for achievement that powered his life.

I

A Glimpse at
Procter's Character

When William Procter, Jr., died in February 1874, an era in American pharmacy came to a close. To the generation of pharmacists that established pharmacy's role in society, Procter had served as a model practitioner; after his death, he became a symbol for the next generation. He exemplified what the individual pharmacist could accomplish as a practitioner, teacher, author, and scientist. As an editor and organizer, Procter had helped to forge a national character for a young and aspiring profession. Unfortunately, few personal items survived Procter aside from a handful of letters. In order to learn about the private side of this man, this chapter relies on the voluminous published record of editorials, reports, and articles, as well as on the recollections of friends, colleagues, and rivals. Although these sources reveal only a veiled glimpse at the private man, they allow us to speculate as to the sources of his life-long pursuit of the progress of American pharmacy.

The story of Procter and his contribution to American pharmacy began in Baltimore, Maryland, on May 3, 1817. William was the ninth child of Isaac and Rebecca Procter, proprietors of a successful hardware store. (The junior was added to distinguish him from his uncle William.) Both active members of the Society of Friends, the Procters planned to raise young William in their faith and provide for a liberal education. When Isaac died from yellow fever in 1820, the Procters fell on hard times that all but eliminated any chance of William attending college. With the hardware business divided up through litigation, Rebecca relied on help from relatives and the Quaker community to raise her family. Facing adversity, she stressed the values of honesty, modesty, and love to her young son.[1]

The same values were taught to young William when he attended a Friends school in Baltimore. Set up in the late eighteenth and early nineteenth centuries, such schools provided a "guarded education" for Quaker children, stressing practical lessons from the Bible, while avoiding distracting subjects such as art and music. For a basic education in the 3 Rs probably no better schools existed in the early nineteenth century.[2]

A boyhood friend of Procter said that their schoolteacher possessed "rare gifts and attainments" and encouraged their individual intellectual interests. Young William delighted in mineralogy and botany, bringing specimens to school to show the other boys. Even at a young age, Procter impressed teachers and classmates with both his studiousness and gentle manner.[3]

Procter left the Friends school in his early teens because of another family difficulty. His older sister's husband, proprietor of a cooper's shop, fell ill and asked young William for aid. Charles Bullock, a close friend of Procter, speculated that his experience as a cooper's assistant initiated Procter's life-long "knowledge of tools and . . . dexterity in the use of them." Young William soon tired of this work, however, and asked his mother if he could learn the drug business in Philadelphia.[4]

Procter had become acquainted with the character of the drug business on a previous visit to Philadelphia. His mother was a close friend of Tabitha Turnpenny, whose son was apprenticed to Henry M. Zollickoffer, an apothecary at the corner of Sixth and Pine in Philadelphia. Joseph Turnpenny quickly befriended Procter and proudly showed off Zollickoffer's establishment. Procter decided that pharmacy was more interesting than barrel making and entered Zollickoffer's shop as an apprentice in 1831 at the age of fourteen. For the next forty years, pharmacy dominated almost all of Procter's waking hours.[5]

By 1831, when Procter entered Zollickoffer's shop, the apprenticeship system had begun to decline in most other areas of skilled work.[6] Increased mechanization, improved transportation, and a larger urban population decreased the profitability of small-scale manufacturing enterprises. The mechanic, rather than the skilled artisan, became the new bulwark of the middle class.[7] To educate this new class of workers (and keep them out of taverns), mechanic's institutes and libraries were established in the larger eastern cities.[8] Professional education, most notably medical schools, also expanded during the first half of the nineteenth century.[9] Pharmaceutical education took a middle ground, with apprenticeship serving as the prime mode of learning, and formal lecture courses as an optional "rounding off" of a young man's education.[10] Throughout Procter's life, no effective statutes required any level of pharmaceutical education. Anyone with sufficient capital and nerve could open a drugstore.[11]

Apprenticeship offered to William a valued introduction to the scientific side of pharmacy. Living with the Zollickoffers, Procter developed the habit of rising early in the morning for a session of what he called "self-culture" before going to work in the shop. At first, this consisted of reading the latest chemistry and physics texts.[12] After finishing the course of studies offered at the Philadelphia College of Pharmacy, Procter added history, biography, and moral philosophy to his personal reading program.[13] Forty years later, Procter remembered fondly his apprenticeship for the learning opportunities it afforded him.[14]

Procter's experience is in contrast to those of other pharmacists of the early nineteenth century who generally recalled their apprenticeships as long hours of drudgery. Besides the usual "wholesome development of muscle through wielding the ponderous pestle, handling the sieves and working the screw press," Procter's apprenticeship allowed him time for scientific investigations.[15] Not only was Zollickoffer's shop laboratory open to Procter's efforts, but just four blocks north sat the drugstore of Elias Durand (1794–1873). Durand had received his pharmacy training in France and had served as a *pharmacien* in the Napoleonic army. After Bonaparte's fall, Durand came to America and eventually established his apothecary shop in 1817. An avid amateur scientist and a leading member of the Academy of Natural Sciences of Philadelphia, Durand made his pharmacy a center for botanical and pharmaceutical experimentation in early nineteenth-century Philadelphia. Procter described Durand's shop as "the daily resort of such men as Drs. Horner, McClellan, Mitchell, Meigs, Mutter, Bache and Goddard."[16] Through his friendship with Durand and his able apprentice, Augustine Duhamel, Procter learned of the great achievements of French pharmacists such as Pierre-Joseph Pelletier, Joseph-Bienaimé Caventou, and Pierre-Jean Robiquet during the 1820s and 1830s. Durand encouraged young Procter to pursue serious investigations and had an expensive set of metric weights imported from France for Procter to use.[17] This interaction probably inspired Procter to obtain the working knowledge of French that he displayed in later years and emulate French pharmaceutical science long after Germans gained ascendancy in the field.[18]

Procter's formal education, broken off several years before, resumed in 1835 when he enrolled at the school run by the Philadelphia College of Pharmacy. Active members of the College, like Zollickoffer, allowed their senior apprentices time off in the evenings to attend lectures. The College set up its course similar to those of medical schools of the day: students attended the same series of lectures twice, wrote a thesis, and passed an examination. To receive the diploma as a Graduate in Pharmacy, the student had to possess four years of apprenticeship with a member of the College. Although the requirements were not particularly difficult, most students opted to attend just some of the lectures and forgo graduation. They attended lectures to round out their education and viewed sitting through the same lectures twice as a waste of time and money. In addition, neither the state nor the public regarded a school diploma as a requirement for practice.[19] Procter, however, fulfilled all requirements and graduated from the school in 1837. Less involved than medical education, which took up about five or six hours of the student's day for three or four months, pharmacy instruction took place two or three evenings a week, after the shops were closed.[20] During the first twenty years or so of the College, two courses of lectures were taught: materia medica and pharmacy, and chemistry.[21] Procter heard these subjects taught by two recognized experts: Joseph Carson and Franklin Bache.

Carson, who had his first training as a pharmacist, graduated from the Medical Department of the University of Pennsylvania in 1830. Taking over the materia medica and pharmacy course at the College's school in 1836, Carson also assumed the editorship of the *American Journal of Pharmacy* published by the College. Author of works on medical botany, Carson succeeded the renowned George B. Wood as professor of materia medica at the University of Pennsylvania in 1850. Procter was fortunate to have Carson as his teacher in this important course. All too often in nineteenth-century pharmacy schools, the combined materia medica and pharmacy course was taught by a physician who cared little about pharmacy and taught less. Carson, with his pharmacy background, was an exception to this general rule.[22]

Even as early as the late 1830s, Franklin Bache (1792–1864) established himself as a prominent medicinal chemist as joint author of the *Dispensatory of the United States* with George B. Wood. In contrast to the fluent lecture style of Carson, Dr. Bache delivered his instruction in a slow, precise fashion, emulated by his student Procter. Together with his close friend Wood, Bache dominated the process of drug standardization in the United States during the mid-nineteenth century.[23]

Carson and Bache provided Procter and his classmates with the best instruction available in the United States in their two areas of expertise. Procter's contact with these two leaders of Philadelphia medicine and pharmacy later opened up important avenues for his career. In 1846 Carson supported Procter's candidacy to become the first occupant of the newly created chair in practical pharmacy at the College; and in 1848 Carson took Procter as his coeditor of the *American Journal of Pharmacy*. Impressed by Procter's abilities as a pharmaceutical chemist, Bache asked him to collaborate in the revision of the *United States Pharmacopoeia* and the *United States Dispensatory,* key medical and pharmaceutical works of the time.

Procter learned his subjects well, and in 1837 passed the examination administered by the College committee and received his "degree of Graduate of Pharmacy."[24] Along with learning the basics of pharmaceutical science, Procter had gained values and ideals that would guide him throughout his career. Although Bache and Carson practiced medicine, they expressed concern with the future of pharmacy practice. A higher level of professional practice by pharmacists promised a more reliable supply of drugs and medicines. Bache emphasized the utility of chemistry in the discovery of new drugs and the improvements of old drug preparations. Both of these outcomes, he hoped, would push back the forces of quackery.[25] Carson tried to inculcate some vocational pride in his students by illustrating the successes of pharmaceutical science, and by downplaying the commercial side of pharmacy.[26]

Procter remained with his preceptor after finishing his apprenticeship and schooling in 1837. As Zollickoffer's senior clerk, he had the pharmaceutical responsibilities of the owner without the financial obligations, a situation that provided him with ample opportunities to hone his skills. With his evenings free from ledger sheets and the burdens of business, Procter continued his self-imposed program of education. In 1840 he attended lectures given by some notable Philadelphia scientists of the day, including the famed chemist Robert Hare. Afterwards, Procter went back to Zollickoffer's and replicated the demonstrations given by the professors that evening.[27] In 1841 he acted as the secretary of the College's committee on the revision of the *Pharmacopoeia,* another valuable experience. During the next two years he learned French and read more literature and history. Finally, in 1844, at the age of 27, Procter left the secure environment of Zollickoffer's shop to set out on his own.

In his diary Procter reflected on what the future held for him in his new shop at Ninth and Lombard:

I am about to leave Sixth and Pine streets, after so long a residence. What singular events occur! Little did such a prospect appear probable some years ago. Steadiness and calmness of mind, how important to the proper appreciation of life! This I daily become more convinced of, and find cause to note the want of it in my own case. Reflection steadily and calmly directed to moral and intellectual improvement, with all the rigor of justice, and all the affection of mercy, how few can truly govern themselves! I have made little progress in this all-important power, and have too frequent cause to regret acts of indiscretion and weakness.[28]

As Rorabaugh has pointed out, the 1830s and 1840s were difficult times for apprentices and young artisans. As skilled trades became more and more mechanized, underemployed and pessimistic young tradesmen turned to hedonistic pursuits and eschewed self-improvement. An entirely new segment of the economy geared toward entertainment and distraction arose to meet the exploding demand. Perhaps Procter had been tempted by a music hall or opera house. For the purpose for his existence, young William had chosen progress based on self-improvement. As prestige and personal achievement came to this quiet, reflective man, he became a strong advocate for the approach that had guided him.

The diary extract is full of self-doubt, but as Procter matured he became more confident in his abilities. Still, he retained the humility and sincerity that had marked him early as an exceptional individual. His diary did not survive, but the published comments of his friends and acquaintances do. They paint a picture of a quiet, slightly introverted gentleman with a sincere belief in the progress of his calling through individual improvement.[29]

On the occasion of his death, several men described the personality and character of William Procter, Jr. A few words recur: moral, sincere, sober, quiet, genial, and most of all, modest.[30] Evan T. Ellis, a Philadelphia pharmacist who saw Procter regularly, remarked that his "modesty and retiring nature" never changed, even as his fame grew.[31] As one colleague put it, Procter was "[m]odest and diffident even to the extent of rendering injustice to himself."[32]

Procter's modesty and shyness made him reticent with strangers, especially groups. When Frederick Hoffmann, a fellow pharmaceutical journalist, first met Procter, he was surprised that such an expressive writer spoke so quietly and with so much reserve. Away from crowded association meetings or hotel lobbies, Procter became "a genial and sympathetic companion and friend."[33] Former students commented on his readiness to answer any pharmaceutical question, any time.[34]

For Procter this kind of help usually took the form of listening as much as talking. At pharmaceutical meetings, Procter contributed to discussions by first sitting back and absorbing all the points of a controversy. After the initial heat of debate subsided, he would take the floor and summarize the alternatives before the group. This approach extended to personal conversations, during which he often would listen and then organize the facts as he saw them.[35] His simple and direct style elicited confidence in those who talked with him, so it is no surprise that he was a much-sought confidant and adviser.[36]

In contrast with his readiness to counsel, Procter had little taste for elaborate merriment. After the organization of the American Pharmaceutical Association in 1852, he repeatedly objected to the lavish banquets and excursions that became regular affairs at the annual meetings of the association. To Procter, they wasted money and distracted members from the important issues at hand.[37] Procter did not object to social gatherings per se. As he said in 1860, "it is more the manner [of the banquet] than the thing itself; and when stripped of vinic demonstrations and alcoholic effusions, 'annual dinners,' if kept within the bounds of moderation, with less show and more substance, may really be the means of promoting friendship and good fellowship among the members."[38]

In the rambunctious days before the Civil War, such opinions did not sit well with many members of the association.[39] Efforts by Procter and others to limit festivities at meetings elicited some spirited responses. For example, John Meakim, a prominent figure in New York pharmacy, responded to those "respected associates . . . averse to any allotment of time for the social discussion of subjects" by remarking: "[W]e have endeavored that our social intercourse on this occasion shall partake of a spirit of investigation, and with that view we should invite your attention to an examination into the science of Gastronomy in its relations to the Hygienic art, with which that of Pharmacy is so closely connected."[40] In matters of professional policy, Procter's views usually carried the day, but his social conservatism was out of step with the common association man.

For Procter, association meetings, like almost every waking hour, should be dedicated to work. As illustrated by the diary extract above, Procter viewed self-improvement through industry as a high virtue. After he died, friends, students, and colleagues commented on his strong work ethic. Perhaps a little bitter, his second wife, Catherine, reflected, "how his *life* was given to the work! how his *heart* and *soul* were in the work! his own personal interests being as nothing in comparison with those of the Philadelphia College of Pharmacy, and with the advancement of the profession."[41] Joseph Remington, a Procter student and leading figure of American pharmacy in the late nineteenth century, put it well when he called Procter "a genius, if by this is meant the capacity for great labor."[42] The twenty volumes of the *American Journal of Pharmacy* best exemplify Procter's ability to put in long hours of work. Procter reviewed scores of medical and pharmaceutical journals and books. Upon receiving a new edition of a medical dictionary, for instance, he would go through all the entries and publish a long list of errata in his editorial column.[43]

Although almost single-mindedly dedicated to the progress of pharmacy, Procter did have two major diversions. The first was farming. In 1855 he bought some land in Mount Holly, a small community across the Delaware River in New Jersey. There Procter and his family would go to get away from the heat of Philadelphia in the summer and spend time tending fruit trees.[44] As mentioned before, Procter had always enjoyed going on excursions for plant or rock specimens.[45] At Mount Holly he could do this and leave the worries of Philadelphia behind. Holding enough land to cultivate was highly desired in mid-nineteenth-century America. Intellectuals enjoyed the challenge of making nature obey their orders. By raising fruit trees, a respectable man like Procter could both recreate and uphold the work ethic.[46]

Travel was the second of Procter's diversions from pharmacy. Notebooks examined by Bullock recount journeys taken after graduating from pharmacy school in 1837 and before beginning his own business in 1844. These trips included visits to Washington, D.C., and the Potomac region, Cleveland and Niagara Falls, and Boston. On his way home from Niagara he passed through upstate New York.[47] Impressed with the beauty of the region and its interesting inhabitants, Procter traveled through upstate New York often in later years when going to or returning from western meetings of the American Pharmaceutical Association.[48] While traveling, Procter kept a notebook full of information concerning the local flora and fauna. Geology, a childhood love and fashionable scientific pursuit, filled his pages. Local pharmacy practices, although mentioned, received less attention.[49]

Once established in his own Philadelphia shop, Procter found it more difficult to obtain time for travel. Association meetings and assignments, however, provided excuses for leaving home. In 1853 he called on the secretary of the treasury on behalf of the association and took the opportunity to visit the Smithsonian Institution and its director, Joseph Henry.[50] Procter's affection for travel, in addition to his dedication to organized pharmacy, might help explain why he missed only one association meeting from 1852 to his death in 1874. These meetings must have been the only journeys he could justify to his family and his own business concerns.[51]

On his return to Philadelphia from a trip, Procter filled his editorial columns with descriptions of the meeting, often adding notes on his own personal travels.[52] For example, after the 1854 association meeting in Cincinnati, he wrote of a new variety of flaxseed that grew in the interesting countryside near that city. A hardy traveler, Procter failed to mention that he was the only Philadelphian who ventured across Pennsylvania and Ohio in the midst of a cholera epidemic.[53]

In April 1867 Procter left the United States for the only time to visit Europe and represent the American Pharmaceutical Association at the Second International Pharmaceutical Congress in Paris. Gone for six months, Procter came back changed; he had lost some vigor but gained a beard. Before the trip he had strongly opposed any governmental regulation of pharmacy; on his return to Philadelphia he supported pharmacy laws.

In Europe Procter took the grand tour, visiting France, Italy, Austria, Germany, and the Low Countries. Along the way he observed both the nature of the land and the pharmacy practiced in each locality. He met with several pharmaceutical and chemical luminaries of the time: Theophilus Redwood, Daniel Hanbury, and John Attfield in England, François Dorvault in France, Justus von Liebig in Germany, and Anton Schürer von Waldheim and Friedrich Flueckiger at the congress.[54]

With all the pharmaceutical opportunities open to him, Procter wrote home that "it should not be expected that the traveller, albeit a pharmaceutist, (urged on his journey by the every recurring necessity of seeing much in a short time,) should linger among his specialities" and referred to himself as "temporarily emancipated from the pestle, wandering amid the olive shades, the orange groves and vineyards of Southern Europe, . . . not tempted to view them from the standpoint of the Pharmacopoeia." Such a statement, however, shows how much pharmacy was on his mind at all times, enough so to make him feel guilty for his actions.[55]

Procter went out of his way to see the great sights, natural or man-made. Hearing that the overland journey from Nice to Genoa on the Corneche road held special beauty, he chose that bone-jarring route rather than the comfortable steamer ride.[56] When in Paris he enjoyed not only his trips to museums and the exposition, but a tour of the great sewers of that city.[57] Above all, though, the botanical wonders of Europe impressed him. In the *Journal,* Procter reported in detail on the agriculture and natural flora of that part of southern Europe through which he passed.[58] On his return Procter wrote that perhaps the most impressive sight was the magnificent botanical garden at Kew near London.[59]

Procter came home with a different view of continental pharmacy than he had carried across the Atlantic in April. He had expected the International Pharmaceutical Congress to be an impressive forum for the exchange of pharmaceutical knowledge; instead he found delegates haggling over political issues. When pharmacy in the United States came under attack at the congress as an example of the shortcomings of nonregulated practice, Procter responded with a short address. Instead of defending the open competition that existed in American medical and pharmaceutical practice, he described the progress made in the previous fifty years and the prospects for the future. Procter kept his own personal opinions largely to himself and stuck to the facts as he saw them. In the midst of his address, however, Procter digressed to state that he favored a system that allowed only graduates of pharmacy schools to practice pharmacy, although such practice should be subject to open competition. During his travels on the Continent he concluded that governmental interference could benefit both the public and pharmacists. Even though the European trip shattered some of his ideals concerning continental pharmacy, it reinforced his resolve to work for the progress of pharmacy in America.[60]

After his European trip Procter confined his journeys to annual meetings of the association. His weakening body might have dictated this decision. When traveling in the Alps, Procter had insisted on going up the highest peaks, even though he suffered from heart disease. His traveling companion, Albert Ebert of Chicago, recalled: "I remember one occasion particularly when [we] were ascending the Wenger Alps. We had started early in the morning, and had ascended some three thousand feet when Prof. Procter collapsed. . . . [F]or half an hour we thought he would surely die. We used the limited restoratives we had, and after he recovered a little he pressed his hand to his heart, and when he could speak he said: 'I have heart trouble; I ought not to have made this ascent.' That was the first we knew of this trouble."[61]

Procter's heart condition may have been a life-long problem.[62] In 1853, for instance, Procter mentioned preparing "a variety of diuretics, for a member of my family affected with valvular disease, attended by dropsical effusion into the pleura and pericardium." Did he make it for his first wife, Margaretta, or for himself?[63] In 1866 Procter resigned from his professorship because of his health, but there seems to be some evidence that the heart attack in the Alps reduced his vigor even more.[64] After 1867 Procter resisted scheduling association meetings during periods of hot weather, which aggravated his condition. In addition, when the association published a portrait of Procter in its *Proceedings* following his death, the editors purposely chose a pre-1867 pose, when he was "in his full vigor."[65]

When the College's professor of pharmacy, Edward Parrish, died in 1872, Procter returned to teaching.[66] His assistant at the time, Joseph P. Remington, remembered how Procter's "heart

The bearded visage in this photograph of 1867 is the way
subsequent generations of pharmacists mostly remembered Procter, even though he was clean shaven until
about seven years before his death. This image served as
the basis for an oil painting now hanging in the Joseph
England Library of the Philadelphia College of Pharmacy
and Science. (From Kremers Reference Files, University
of Wisconsin–Madison School of Pharmacy, Madison,
Wisc.)

trouble . . . was rapidly undermining his strength, and I found him more than once on the landing of the college steps hesitating about mounting a dozen steps more to reach his lecture room." In early 1874 Procter decided to resign from his chair at the end of the term. On February 9, after giving one of the last scheduled lectures of the year, Procter went home, talked with his wife until late, retired for the evening, and soon after died of a heart attack.[67] Fully cognizant of his condition, he had continued to work in his pharmacy and teach at the school.

Some of Procter's inner strength can be attributed to his upbringing in the Quaker community of Baltimore and his continued participation in Friends activities in Philadelphia.[68] Although outspoken in his belief in the progress of the pharmaceutical art and science, Procter made "but little profession" of his faith as a Friend. Still, he observed the Sabbath and regularly attended the Spruce Street Monthly Meeting in Philadelphia.[69] Procter's belief in God showed in his writings, but only rarely. Subscribing to the idea that each man had to find his own God, Procter concentrated his efforts on practical matters at hand, avoiding dogmatism in either religion or pharmaceutical matters.[70]

In contrast to his religious beliefs, Procter's personal heroes are obvious from his writings. Soon after beginning his editorial column in 1850, Procter included obituaries of leading figures in pharmacy.[71] Some of the men Procter admired, such as the English professor of materia medica Jonathan Pereira (1804–53)[72] or the eminent French pharmaceutical chemist Eugène Soubeiran (1797–1858), he never met.[73] Others, like Elias Durand and Procter's older brother Stephen, were intimate role models. William had followed Stephen's footsteps into pharmacy.[74] At Durand's pharmacy, Procter began his career as a pharmaceutical experimenter and gained a friend, Durand's apprentice and later partner, Augustine Duhamel.[75] Four years older than Procter, Duhamel probably served as an older brother to William during the 1830s and early 1840s. His death in 1846 struck Procter hard.[76]

As an American of English descent, Procter felt a close connection with British pharmacy.[77] As editor of the *Journal* he corresponded with some of the leading figures of British pharmacy, such as Jacob Bell and Daniel Hanbury. Procter respected Bell for his role in organizing the Pharmaceutical Society of Great Britain and his dedication to public service.[78] He also admired Bell's "practical good sense in dealing with the pleasures, the duties and the difficulties of life."[79] Though Procter never met Bell, he did meet Hanbury during his trip to Europe. They had corresponded during the 1850s and 1860s, mainly about botanical materia medica.[80]

Although influenced by the example of others, Procter was his own man. He matured during a period of American history when an individual possessed the freedom to try out new and bold ideas. American science, long burdened by its debt to Europeans, sought to exhibit its own discoveries and achievements. A new generation of idealists began to question the old economic and social order that had produced the panic of 1837 and the controversies over the Bank of the United States. Instead of just seeking personal wealth, the generation born after the War of 1812 looked to reform and progress as the way to a better future.[81] Procter participated wholeheartedly in this pursuit of progress, even though he limited his efforts to pharmacy.

The idea of progress "owed much of its popularity to the amazing growth of the West and the startling achievements of science and technology. By the 1840s it interpenetrated every aspect of American thought and feeling."[82] Of course, the idea of progress had achieved some currency during the Enlightenment, but the practical successes of science and the dreams of Romanticism energized this nineteenth-century version.[83] Procter occasionally expressed "amazement" at the progress he saw around him, like the telegraph, new railways, or transatlantic steamers.[84]

Exactly what was this idea of progress? It "embraced the traditional concept of advance and melioration. In contrast to a violent or revolutionary change, progress was considered to involve a regular and gradual process of growth. The concept of intellectual and moral improvement was used to distinguish the idea from those mere material and physical advances which might involve a change without true individual and social betterment."[85] Procter, on the other hand, did not separate so closely intellectual and material advances, as he indicated in 1848:

The important influence exercized on the arts for their improvement, by the discoveries in chemistry which are so constantly coming to light, is the cause of deep satisfaction to all who give a thought to the progress and amelioration of our race, through the increased facilities they afford to an enlightened civilization. Such, especially, are those discoveries which tend to cheapen and increase the production of substances closely connected with the comfort of mankind, and upon which all depend. He, then is a true benefactor to his fellow men, who, whilst immured in the recesses of the laboratory, closely interrogating nature, elicits from her revelations fraught with mighty consequences to the economical relations of society.[86]

Procter expressed enthusiasm for American progress, advocating the exploitation of natural resources and the public support of technical education.[87]

A conservative man, Procter supported steady progress, but not "wild-eyed" reform. Before his trip to Europe, Procter rejected any suggested reforms or laws that could infringe on open competition or give too much power to any government. Like many Americans, Procter suspected that many, if not most, reformers were self-serving.[88] He shared with Edward Parrish the belief that progress in American pharmacy could only be obtained through open interchange of information and conscious self-improvement.[89] The extract from Procter's diary showed his dedication to self-improvement, a value that attained widespread favor in the mid-nineteenth century.[90] To Procter, meaningful self-improvement could only be obtained through self-sacrifice. Thus Procter avoided the reading of novels and other pastimes he considered unproductive.[91]

Yet Procter did not lead an isolated life. He was keenly interested in national and international affairs, albeit with a pharmaceutical slant. He particularly studied British history because American pharmacy arose from a transposition of English pharmaceutical practice during the colonial period.[92] At the same time Procter occasionally expressed fierce nationalism when discussing scientific or pharmaceutical affairs. The United States had much to offer the world, in his opinion, if Americans would work hard and Europeans would recognize their efforts.[93]

Procter accomplished most of his own hard work in his shop at the corner of Ninth and Lombard. There he practiced exceptional pharmacy. He did not stock patent medicines or casually dispense poisons, as did most pharmacists. Instead Procter based his practice on precision and service to the medical and pharmaceutical community as exemplified by drug assay or the preparation of unusual medicines. Happy in his small domain on Philadelphia's south side, Procter chose not to build a large business on his reputation. Instead of dedicating much of his drugstore to the sale of fancy goods or to a soda fountain, as was commonly done, Procter expanded his laboratory and added a small writing area.[94] At this desk Procter composed the scientific papers, practical articles, editorials, and lectures upon which his reputation eventually grew.

In contrast with the published record of his pharmaceutical achievements, only a few descriptions of his personal life survive. In October 1849, five years after starting his own shop, Procter married Margaretta Bullock at Mount Holly, New Jersey. Together they had two children, Wallace (b. 1851) and Mary (b. 1852). In 1859 Margaretta passed away, leaving William with the two young children.[95] Procter remarried in 1864, to Catherine Parry, also of New Jersey.[96] They had no children, but evidently a happy marriage. Catherine almost always went with William to association meetings, where she was well known among members.[97] She survived William by thirty years, dying in 1905.[98]

Procter's only son, Wallace, died soon after, in 1911. He had followed his father into pharmacy, attended the Philadelphia College of Pharmacy, and graduated first in his class. After his father's death in 1874, Wallace carried on the family business for fifteen years, but eventually moved from Philadelphia to work for the Ohio Valley Drug Company in Wheeling, West Virginia.[99]

Procter had composed his will in April 1867 before crossing the Atlantic en route to the International Pharmaceutical Congress.

Although not filled with the long stipulations of some wills, Procter's does indicate the important people in his life.[100] He also made provisions for allowing his clerk, David Preston, to buy a full partnership in the shop at Lombard and Ninth from Wallace.[101]

On February 11, 1874, the Philadelphia *Public Ledger* carried the following obituary: "PROCTER—Suddenly on Third day morning, the 10th inst. Professor WILLIAM PROCTER, Jr. in the 75th year of his age. Funeral from his late residence, S.W. corner of Ninth and Lombard streets, on sixth-day morning, at 8½ o'clock. Interment at Mount Holly."[102] The cause of death, according to the attending physician, was "angina pectoris."[103] A postmortem examination "disclosed considerable ossification of the aorta."[104]

On a cold Friday, February 13, delegations from pharmacy associations along the East Coast came to the funeral. His last class at the college arrived as a group and followed the funeral procession to the Market Street wharf. There they surrounded his casket with flowers and watched it cross the Delaware to New Jersey.[105] From Camden it was taken to Mount Holly, where Procter was buried next to Margaretta, in sight of his beloved country home.[106]

Procter and the progress of American pharmacy went hand in hand for forty years. Out of the hours of self-study and slow time in his shop Procter formulated an unwritten program to further his cause. For American pharmacy to continue its professional development, to build on the advances made during the 1820s and 1830s, Procter advocated ethical pharmaceutical practice, active scientific investigation by practitioners, general acceptance of uniform drug standards, an aggressive professional press, local schools of pharmacy, and cooperative effort.

In each of these areas, Procter made a mark, sometimes setting the course for that aspect of American pharmacy for decades to follow. As we will see in the following chapters, however, Procter's legacy was mixed. Most American pharmacists were not modest, self-motivators interested in science—they were small shopkeepers who had ambitions for financial success and public esteem. The institutions created or influenced significantly by Procter's ideal of professional progress based on individual self-improvement, so well exemplified by his own contributions, largely failed to keep up with the changes brought on by developments outside the profession.

As a practitioner, however, Procter did serve as an excellent role model, one that had continuing influence well into the mid-twentieth-century. A look at his manner of practice tells us much about how American pharmacy changed during the middle half of the nineteenth century.

William Procter as Practicing Pharmacist

2

In May 1844, Procter left the secure environment of Zollickoffer's home and shop to start his own pharmacy. Proud of his education and experience, he felt confident that he could succeed. He planned to combine the best qualities of Zollickoffer's retail business with Durand's laboratory science. This would put his shop in stark contrast to a common apothecary's shop in Philadelphia, which was a dusty little establishment without the proper equipment and a full complement of basic ingredients. Instead of making the compounding of medicines a sideline as did most apothecaries, Procter would make it the centerpiece of his business. The ethical and scientific standards he learned at the College of Pharmacy would guide him in practice and set him apart from those who ignorantly sold drugs and poisons. A few months later, after a discouraging beginning, Procter wrote: "It has been a time of singular discomfort to me, the anxiety incident to opening a new store, and the much time unemployed has been very burdensome. I need more faith and confidence in the course of events." He had no idea that his business, struggling for several years, would provide him with an opportunity to "employ" himself in many different ways.[1]

When Procter opened his shop at the southwest corner of Ninth and Lombard, the last thing Philadelphia needed was another drugstore. Competition pressed hard, with no legal restrictions covering who could or could not practice pharmacy.[2] Americans relied on a free marketplace to regulate itself.[3] Prescription-oriented retail pharmacists, such as Procter, had just arrived on the American scene, achieving some sort of self-identity only in the previous thirty years or so. In the 1840s they still practiced mainly in the largest cities and towns where some physicians wrote prescriptions. Other physicians generally dispensed their own medicines from kits and chests prepared by wholesale druggists. In smaller cities, men like Procter competed not only with other pharmacists, but also with physicians who ran their own shops and with general stores that often carried drugs and medicines.[4]

The competition with the general stores was particularly significant in the nineteenth century, when pharmacists received a small part of their livelihood from the compounding of prescriptions. Sundries, such as cosmetics, toiletries, dyes, small housewares, and other inexpensive goods, kept many a pharmacist in business. Aside from these common products, some drugstores carried a selection of books or, especially in rural areas, paints and oils. Even those pharmacists of the mid-nineteenth century who wished to concentrate most of their business in pharmaceutical areas stocked these.[5]

Nostrums were a different matter. Sold by many retail stores in mid-nineteenth-century America, nostrums presented a pharmacist with a dilemma. They provided needed income, but jeopardized his prescription practice, the basis of his occupational aspirations. In addition, since the formulation of a nostrum was unknown, the pharmacist could not easily offer his own competitive product.[6] Even avid advocates of strict pharmaceutical practices found it difficult to condemn pharmacists for stocking them. Yet Procter did not stock them and spoke strongly against their sale.[7]

When Procter opened his drugstore, he faced not only stiff competition from other Philadelphia shops, but a tough neighborhood surrounding him. Situated near the south border of the city, an area frequented by gangs of young hoodlums, Procter's shop stood little chance of attracting a genteel clientele. Many of the lots in this neighborhood, including the whole square block catercorner from the store, were empty. Only after the police cleaned up the area did more people set up homes there and patronize Procter's establishment.[8]

During the 1840s, with his business slow, Procter could take time for other things. He turned some of his efforts toward scientific work, continuing those projects he had begun as an apprentice and clerk with Zollickoffer. The process of displacement and the qualities of volatile oils attracted his interest.[9] Problems at the College also drew his attention, and in 1845 Procter, Duhamel, and Edward Parrish presented a proposal to a College meeting calling for the establishment of a chair in practical and theoretical pharmacy.[10] Almost a year later, in June 1846, the Board of Trustees of the College unanimously elected Procter to the new professorship. Although student fees did not amount to a large sum of money, they helped, and the professorship tended to increase Procter's stature among the clientele of his pharmacy.[11] The *American Journal of Pharmacy* provided a supplemental income after 1850, when Procter assumed the editorship of the *Journal* and received a small annual stipend for his work.[12] Soon afterward, in 1851, Procter traveled to New York for a convention of pharmacists that led to the formation of the American Pharmaceutical Association the next year. These outside activities may have been encouraged, rather than inhibited, by the poor state of Procter's business in the 1840s and early 1850s. If he had been busy from the start, he might not have devoted himself so unreservedly to these broader pharmaceutical concerns.[13]

In his early years at Ninth and Lombard, Procter watched in frustration as prescriptions went to competitors. Standing behind his counter waiting for challenging requests, he received only five or six prescriptions per day, often for the most mundane of preparations. On February 27, 1847, for instance, he filled six prescriptions, for which he charged a total of about $1.20.[14]

The first was from a Dr. Mayer, who wanted fifteen grains of camphor dissolved in one-half ounce of ether, probably for his own use.[15] The second prescription was for a Mrs. Clymer. The ingredients suggest that a tonic was being given following a bout of intestinal troubles: ammonium hydrochloride, and syrups of ipecac, wild cherry, and balsam of Tolu, flavored with orange flower water.[16] Procter asked fifty cents for this mixture, close to the maximum he charged for a single prescription.[17]

The third prescription Procter compounded called for a mixture of Dover's powder and potassium nitrate, divided into six powder papers. Although "very little art" was involved in such a prescription, Procter charged nineteen cents for it, because of its expensive ingredients (opium and ipecacuanha).[18] The next request, marked no. 326 by Procter, just asked for an ounce of uva-ursi or bearberry. For this botanical drug used to treat urinary tract disorders, Procter asked four cents. Physicians commonly ordered simple ingredients like this from which a nurse or a family member would make an infusion or decoction.[19]

After this modest order, Procter received a fairly typical prescription, which called for the combining of two crude botanicals with two chemicals and their subsequent division into powder papers. Again, its ingredients (American columbo, valerian, an iron salt, and bicarbonate of soda) and its dose (three times a day) suggest a stomach tonic.[20]

Procter's final prescription of February 1847 called for zinc acetate to be added to a base of cerate of carrot. Procter asked twenty-five cents for this mildly astringent external application. The time and effort that went into making the base probably accounted for the relatively high cost. In later years "Procter's carrot ointment" (actually a cerate) became well known among Philadelphia physicians.[21]

The prescriptions discussed above indicate that Procter did not spend a great deal of his time each day filling such orders, for most did not call for special effort. In addition, it shows a low income from pharmaceutical practice. Once raw material costs are discounted, Procter probably cleared about one-half dollar a day. This was less than an unskilled laborer earned (one dollar a day).[22] Because of his financial situation, Procter may have been forced to look for outside sources of income, such as teaching and editing. Perhaps a more challenging pharmaceutical practice would have kept Procter away from his own scientific experiments and limited his contribution to the literature.

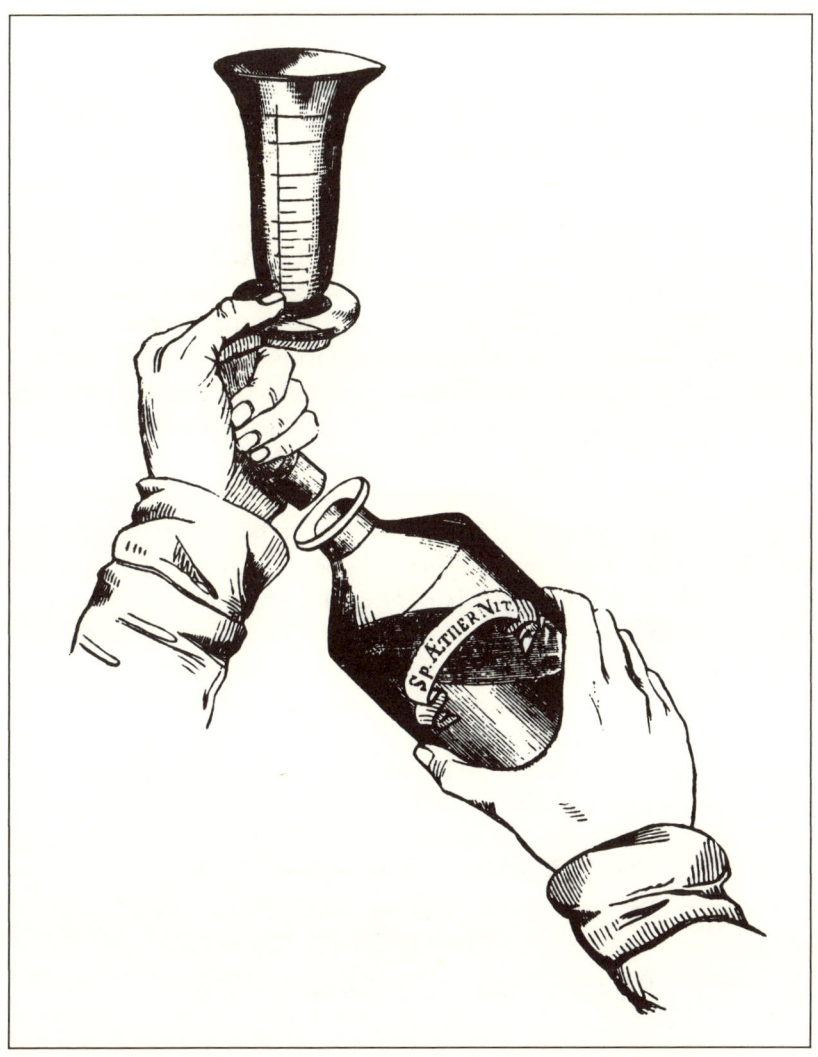

Practical Pharmacy, by Mohr, Redwood, and Procter, provided illustrations showing the basic details of pharmaceutical practice. Here is the proper method for holding a measure, just before the pouring of a liquid. (From Francis Mohr, Theophilus Redwood, and William Procter, Jr., *Practical Pharmacy* [Philadelphia: Lea and Blanchard, 1849], 49.)

This small sample does not, of course, give a complete picture of the character of the prescriptions Procter did fill in the late 1840s. An examination of 400-odd prescriptions reveals that ten drugs appeared often: mercury or its salts, opium or morphine, hyoscyamus, ipecacuanha, rhubarb, iodine, antimony, quinine, lead, and squill. The chemicals called for were usually metallic salts or alkaloids, for example, ferrous sulfate or morphine, whereas the botanicals were usually unprocessed, ground crudes. Procter combined these and other drugs into various dosage forms, most commonly solutions, mixtures, and powder papers. To a lesser extent he dispensed pills, plasters, and ointments.[23]

Procter's prescription book also documents the transitional state of American therapeutics at midcentury. During the middle third of the century, physicians gradually set aside the drastic depletive therapies of the previous generation (bleeding, puking, sweating, blistering, and purging), substituting a milder, less aggressive approach. Historians have suggested several reasons for this shift. Some physicians theorized that the general nature of disease had changed from overexciting to debilitating the bodies of the afflicted, which called for supportive rather than depletive therapies. Another group of doctors argued that the "healing power of nature" accomplished better results if left alone. Lastly, homeopaths and other sectarians offered their patients less harsh (and therefore more attractive) therapies than the "heroic" treatments popular among orthodox physicians in the early years of the nineteenth century. Even as this shift in method took place, physicians retained an age-old objective for their therapy: to restore the natural balance within the patient, thereby helping recovery.[24] The movement from depletive to supportive or palliative therapeutics was not abrupt largely because the ultimate mission of the medicines had not changed.

The shift is obvious in the Procter prescriptions of 1847. Specifically, the use of severe emetics and cathartics had declined, and blistering was becoming much less common. Cantharides, or Spanish fly, which was the main blistering agent of the period, was almost absent from Procter's prescriptions. Ipecacuanha, a popular emetic of the previous generation, was prescribed usually in much lower, "tonic" doses. Shifting away from harsh depletive medicines, physicians turned more and more after midcentury to tonics, often alcohol-laden. The idea was to strengthen the "vital power" within the patient, thereby restoring the natural balance within the body. Quinine, ipecac, and other bitter drugs were common ingredients in tonic formulations. (By the 1860s the stimulant of choice became simple beverage alcohol.)[25]

Even though therapeutics had moderated by midcentury, physicians retained many of the old medicines, if perhaps in lower doses. Traditions were not easily set aside, and physicians were ready to bring out the lancet or strong drugs if confronted with the appropriate disease state. Moreover, patients still expected drugs to "work," that is, to have a perceived effect. The presence of the moderately strong laxative rhubarb in many of Procter's prescriptions shows that occasional catharsis remained a common approach. And opium, a bulwark of the materia medica for millennia, had become even more prominent as physicians sought to support and soothe rather than deplete the afflicted patient.

Next to opium and its derivatives, mercury and its salt calomel (mercurous chloride) were the most commonly dispensed drugs from Procter's pharmacy. Along with antimony, iodine, arsenic, and a host of other drugs, calomel was classified as an "alterative." This class played a central role in depletive therapy and continued to be used well into the late 1800s in lower doses. In theory these drugs altered "the fundamental balance of forces and substances which constituted the body's ultimate reality."[26] Calomel was a favorite alterative because it could be used, depending on dosage and duration of therapy, as a laxative, as a stimulant, as a sedative, or as a tonic.[27]

Procter's prescriptions of 1847 are quite similar in nature to those from the Allinson pharmacy in nearby Burlington, New Jersey, which were analyzed by Cowen, King, and Lordi in 1981. The Allinson prescriptions dated from 1854 and also called for a great deal of calomel, opium, ipecac, and quinine. As with the Procter prescriptions, those from Allinson's were usually simple formulas containing one or two ingredients. That there is little difference between the prescriptions of Procter and Allinson is

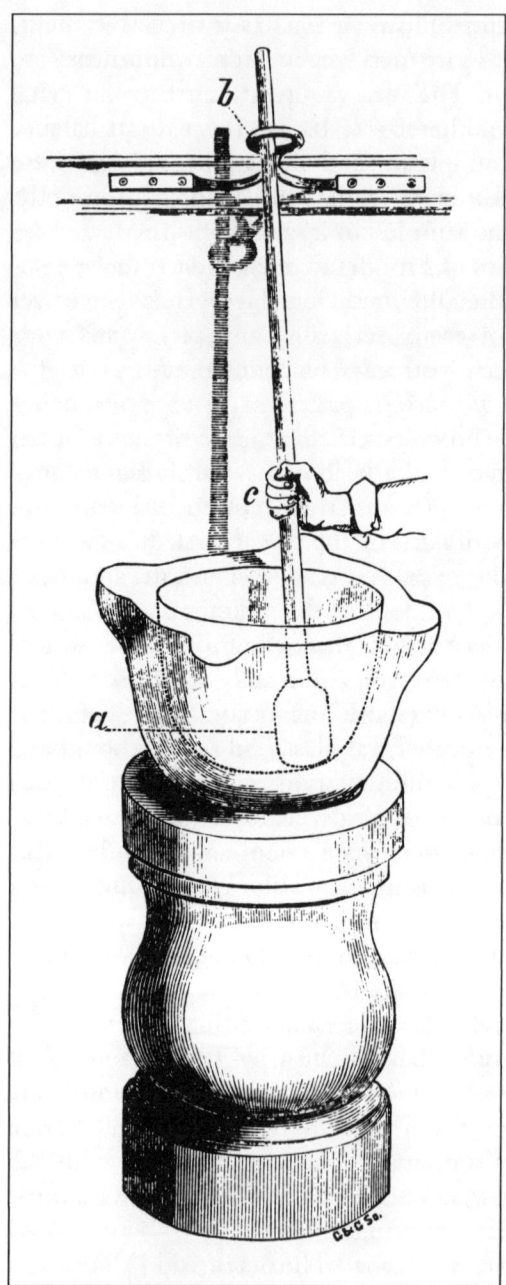

Trituration was a serious matter in nineteenth-century pharmacy. This arrangement of a marble mortar with a hard wood pestle was especially effective with crystalline drugs. It was also used to make mercurial ointment. (From Francis Mohr, Theophilus Redwood, and William Procter, Jr., *Practical Pharmacy* [Philadelphia: Lea and Blanchard, 1849], 153, 416.)

not surprising, for almost all of Allinson's were written by Philadelphia-trained physicians and Allinson's professional practice was small, like Procter's.[28]

Procter's many additions to the book *Practical Pharmacy* help answer some questions on how he practiced pharmacy about 1850.[29] How then was Procter's shop most likely outfitted? Heat came from a coal stove, which had a removable top that could serve as a sand bath or be replaced with a still. For small shops of that time, like Procter's, this stove also served as a laboratory furnace.[30] Around Procter at his dispensing counter were shelves holding bottles of medium size (four ounces to one quart), in alternate rows of liquids and solids. Procter objected to using larger containers because their contents often deteriorated before they were completely consumed.[31] On the tops of the shelving fixtures, Procter placed tin canisters, stacked three or four high, containing herbs.[32]

Adjacent to his dispensing counter, Procter probably had "a shallow upright glass case . . . containing one hundred or one hundred fifty glass-stoppered vials, of capacity from two ounces to half an ounce, arranged in steps, and distinctly and appropriately labelled. These bottles . . . contain[ed] organic alkalies and their salts, neutral organic principles, . . . powerful vegetable products, . . . and a long list of vegetable powders used in prescriptions occasionally, and . . . kept only in small quantity, those of leaves being protected from light." Procter advocated this arrangement, not only to save time, but because one could easily recheck that the proper ingredients went into each prescription.[33]

Serious about drug assay, Procter protected his balance, probably made by Duffy of Philadelphia, with a hinged balance case.[34] Inasmuch as he filled only a few prescriptions daily, Procter used gas and alcohol lamps as a heat source much of the time. Both had certain advantages: alcohol lamps possessed portability, whereas gas lamps better served the needs of analytical chemistry, such as blowpipe work.[35]

For the grinding of drugs, Procter liked Swift's drug mill, a popular apparatus in American pharmacies of the mid-nineteenth century. Similar in design to the common coffee grinder, this mill produced powders of a uniformity necessary for proper execution of certain pharmaceutical processes, like displacement.[36] Finely powdered chemicals often were produced through precipitation.[37]

Even though the prescriptions compounded by Procter were usually routine, the production of their ingredients sometimes presented difficult challenges. The strength of a tincture, for instance, depended on the quality of the crude drug, how well it was garbled before maceration, and whether traditional methods or newer displacement techniques were used to extract the hydroalcoholic product.[38] Many of the vehicles, such as cerates, plasters, and certain medicated syrups, were not especially difficult to prepare, but required time. Mercurial ointment, a curse to every pharmacy apprentice of the mid-nineteenth century, took hours of labor to make.[39] Most frustrating, perhaps, was the tendency of many of Philadelphia's physicians to prescribe ingredients from foreign pharmacopoeias, sometimes in French or German. To fill these prescriptions, a pharmacist like Procter had to maintain a good reference library containing works such as the "French Codex, Jourdan's Universal Pharmacopoeia, Gray's Supplement, Soubeiran's or Guibourt's Pharmacy, and Pereira's Materia Medica," in addition to the *United States Pharmacopoeia* and *United States Dispensatory.*[40]

In the 1850s, as both South Philadelphia and Procter's reputation improved, the small shop at Ninth and Lombard became busier. Yet Procter did not try to cash in on his new-found repute. He claimed to remember all too well how the wide reputation of other more established shops inhibited the growth of his own business.[41] Instead of expanding his shop or founding a manufacturing laboratory, Procter remained in his small shop for thirty years. Procter did enlarge his pharmaceutical laboratory in 1861 and the dispensing area in 1870, but he kept the basic layout of the drugstore pretty much as he set it up in 1844. The prescription counter stayed up front near the door, rather than in the back of the shop. Procter did not have a soda fountain, that ubiquitous contrivance that displaced the prescription counter from the front of most shops after the mid-nineteenth century.[42]

By 1864 the pressure of Procter's various outside commitments forced him to hire a full-time drug clerk, David Preston.[43] Preston had studied pharmacy under Procter at the College and shown himself to be an exceptional student. He soon gained the full confidence of his employer, which allowed Procter freely to attend association meetings and travel to Europe in 1867.[44]

As his business grew, Procter and his assistants did more extemporaneous compounding, which commonly included "pill making, plaster spreading, [and] the preparation of suppositories, emulsions, mixtures, capsules and granules." To do this work properly, Procter thought a pharmacist had to "be familiar with the laws and materials of chemistry" and understand "the important processes of solution, evaporation, and distillation, and their use in making tinctures, syrups, fluid extracts, distilled waters and oils."[45] With such skills in hand, he could fulfill almost any request made to him by a physician or customer.[46]

Procter's own practice went beyond average services. As one of the country's best-known pharmaceutical chemists, Procter turned to drug assay as an additional source of income and personal satisfaction. When a new shipment of exotic or precious drugs arrived at one of Philadelphia's large wholesale houses, a sample often found its way to Procter's laboratory. Wholesalers asked Procter to judge the quality of the crudes before accepting a large shipment.[47] Procter relished this work, calling it "a source of pleasure and profit."[48] As a scientist he enjoyed breaking apart a crude drug into readily attainable substituents, such as resin, alkaloid, oil, gum, ash, and so on; as a practitioner, he sought to reduce adulteration in the drug market to protect both his patrons and the fragile reputation of his young calling.[49] Procter often did not need to use the entire sample of a drug, and added the remaining portion to his stock. As the years went by, his shop became well known for its holdings of rare and exotic drugs.[50]

Procter's shop also gained renown as the place where he practiced the model pharmacy that he advocated in his articles and editorials.[51] After trying new methods and apparatus in his shop, he wrote about his experience in the *American Journal of Pharmacy* or reported on it at association meetings.[52] In scores of short articles and in a special series called "Pharmaceutical Notices,"

Procter related tips and hints on how to prepare drugs and medicines.[53] Many of his methods and recipes, like his cucumber
ointment and directions for making fluid extracts, became standard through publication in the *Pharmacopoeia* or the *United States
Dispensatory*.[54] Procter also systematically searched foreign pharmaceutical journals for practical information and published his
selections in a series called "Gleanings."

In addition to specific recommendations for better pharmacy,
Procter expressed his personal views on practice in the *Journal* from
1850 to 1871. As a young man, he called pharmacy "an honorable
. . . pursuit in which any mind may engage without contravening
the best principles of his nature."[55] Even then, however, Procter
realized that men usually became pharmacists for monetary gain.[56]
Through most of the 1850s, he argued that ethical and sound
practice stimulated business, rather than stifled it.[57] As the decade
of the 1850s came to an end with economic crisis and with quackery on the rise, Procter lashed out at those who had "most shame-
fully prostituted to the business of over medication for the sake of
lucre."[58] After twenty years of independent practice, he exhibited
some of his frustrations when he referred to pharmacy as "the
most arduous, self-immolating, ill-appreciated and ill-requited of
professions." He saw little hope of his calling rising above a mere
trade in the near future.[59]

Throughout his career as a pharmacist, Procter expressed such
opinions concerning practice issues through the pages of the
Journal. Nothing attracted more of his attention than the overall
issue of competency and how to ensure it. Pharmacy, in mid-
nineteenth-century America, was caught between those who advocated a totally free market and those who supported "social
protectionism" as a means to reduce destructive competition.[60] At
the start of his career, Procter supported some sort of regulation, at
times comparing American and continental pharmacy.[61] Struggling in his own business largely because of the stiff competition in
the drug trade in Philadelphia, he probably yearned at times for
the controlled environment of the Continent where some states
limited the number of pharmacies. From about 1855 to 1865
Procter's business improved and he became a strong advocate of

open competition in American pharmacy. His trip to Europe in 1867, however, altered his view again and he saw some value in governmental involvement in the regulation of pharmacy.[62]

Poisons and their improper handling created the loudest call from outside of pharmacy for increased regulation of the drug trade. When an incompetent clerk incorrectly dispensed a relatively harmless drug or medicine, usually nothing came of it; when he accidentally filled the same prescription with an overdose of deadly poison, the owner of the store was sometimes hauled into court and found liable. Although small by the standards of the late twentieth century, penalties of a few thousand dollars wiped out an average pharmacist.[63]

The concern over accidental poisonings, among the public at large and medical societies specifically, grew during the 1850s. Moreover, deliberate poisoning, especially with arsenic, became a stylish way to murder an enemy or unwanted mate.[64] Medical men wanted to prohibit by law the sale of certain poisons without a prescription, arguing that pharmacies were a primary conduit for their sale. During most of the 1850s and 1860s, Procter led an unsuccessful battle against such laws, arguing that responsible people should not have to go to a physician in order to purchase rat poison. He thought that better education of the pharmacist combined with a flexible system of recording poison sales offered the best solution. In addition, free market competition would tend to punish offenders with bankruptcy.[65] After personally observing the advantages of regulated practice in Europe, Procter became an advocate of poison control laws.[66]

In this debate, as in others, Procter tempered his concern over public safety with loyalty to his calling. Commenting on a case where a patient died from an incorrectly compounded prescription, for example, he insisted that his readers "not forget the sympathy which is due to the apothecary, who, wounded in his reputation and business, and his deep sensibilities harrowed by the untimely, and too often ungenerous remarks of the unthinking public, finds the nine hundred and ninety-nine instances of careful and highly responsible service have been forgotten, if the thousandth falls short."[67]

Although seemingly limited in their impact, cases of poisoning by faulty compounding continued throughout the mid-nineteenth century and called attention to the wide variation of competency held by American pharmacists.[68] Following the Civil War, some physicians renewed their call for more regulation of pharmacy.[69] Procter, fresh from his trip to Europe, voiced reserved support for laws that attempted to ensure public safety. He did not, however, favor limiting the number of pharmacies, as many European countries did. He claimed that if legislatures eliminated competition, pharmacy would lapse "into a vegetative, red-tape, almost fossilized condition."[70] Instead, Procter favored state laws that would register pharmacists who met certain educational requirements.[71] Not particularly interested in politics or the law, Procter did not take a leading role in the writing of pharmacy law, as did his friend and colleague John M. Maisch (see chap. 7).

Issues that directly concerned daily practice, such as the interactions between pharmacists and physicians, or the quality of drugs, interested Procter more. Because pharmacy was a young vocation in the United States that aspired to professional recognition, its practitioners had to cultivate carefully a good working relationship with physicians.[72] Economic conditions and open competition in both occupations made this a challenging proposition. With the division between medicine and pharmacy unregulated by law and often misunderstood by the public, friction between the two was inevitable. Physicians dispensed, as they had before the emergence of the dispensing pharmacist, and pharmacists prescribed for patients "over the counter," usually because of the poverty of the patient or themselves.[73] Procter, conscious of the damage that "counter prescribing" was having on relations between pharmacists and physicians, condemned the practice from the start of his career. Like other leaders in American pharmacy, Procter emulated the relationship between physicians and pharmacists that existed on the Continent. Above all, he wished to avoid a situation similar to that in England during the seventeenth and eighteenth centuries, where the apothecaries evolved from pharmaceutical into medical practitioners, which led to a long period of interprofessional conflict.[74]

Even when the physician wrote a prescription that was filled by a pharmacist, the prescription itself created problems. Physicians commonly wrote prescriptions in a mixture of Latin (ingredients) and English (directions), which angered curious patients and some secretive pharmacists, who wanted to keep them all in Latin. Procter opposed Latin for secrecy, but hoped that physicians would refer to the *Pharmacopoeia* for the proper name of the drug, and thus encourage standardization.[75]

Because the competency of American physicians varied as much as that of pharmacists in the mid-nineteenth century, educated pharmacists were often presented with dangerous prescriptions written by poorly trained "MDs."[76] Some pharmacists reacted strongly, sending prescriptions containing serious errors back with patients. Procter, always conscious of fragility of the nascent physician–pharmacist relationship, instead advised direct consultation between apothecary and doctor.[77]

In contrast to his friend Edward Parrish, Procter did not propose sweeping reforms for American pharmacy. He hoped that the position of pharmacy could be elevated by the improvement of individual practitioners. For this to take place, Procter insisted that the calling had to be given complete control over itself and its chosen field, the preparation of medicines. In the 1840s and 1850s, physicians posed a serious threat when they dispensed out of their bags or, worse yet, their own shops.[78]

During the late 1850s a new menace to Procter's vision of pharmacy as the independent, science-based profession in charge of medicine making and dispensing arose: large-scale drug manufacturing.[79] Edward R. Squibb, a physician and drug manufacturer, saw the trend as threatening to both pharmacy and his own profession. He observed that "the number of pharmaceutists in the United States . . . who make their own most simple chemical preparations is very small, while [the number] of the physicians who test and examine the preparations they use . . . is perhaps still smaller. The physician relies upon his pharmaceutist, and the pharmaceutist relies upon the manufacturer, and thus the ultimate source of supply is not only removed beyond the confines of the profession, but beyond the reach of professional influence."[80]

Procter agreed with Squibb and worried that large-scale drug manufacturers, driven by business rather than by scientific concerns, would distort the value of their products. Moreover, he feared that manufacturers would substitute their own processes for those outlined in the *Pharmacopoeia,* filling the market with drugs that had the same name but different ingredients or strengths.[81] As will be described in later chapters, Procter tried (and failed) to combat the trend toward large-scale manufacturing through the refinement of in-shop pharmaceutical techniques and the design of new preparations.[82]

After the end of the Civil War pharmaceutical manufacturing burgeoned, causing Procter to comment "that a stand should be made against the encroachments of manufacturing pharmaceutists."[83] As the situation continued to worsen in his view, Procter became more vocal:

When a pharmaceutist makes his own preparations he knows what they are, and is responsible for their quality; he graduates the supply to the demand, and thus renews his stock as often as it is needed. But when once he leaves this true standpoint and abandons his proper business as a *preparer* as well as *dispenser* of medicines, he is at the mercy of circumstances over which his control is very limited. The pharmaceutist who is daily engaged in preparing the medicines he vends, becomes so intimately acquainted with their properties that he can form a fair judgment of their quality when made; but when he forgoes this duty, and depends on the [wholesale] druggist and manufacturer for all the more important preparations of the Pharmacopoeia . . . he cannot trust his sense to the same degree, even supposing he sets out with a supply of good preparations.[84]

Realizing that the position of pharmacists could rise only with increased public confidence, Procter viewed the trend toward large-scale manufacturing as a direct threat to American pharmacy's future.[85]

The federal excise tax placed on alcohol during the Civil War further discouraged pharmacists from making their own preparations. The official methods for producing many extracts involved the evaporation of large amounts of alcohol. When the tax increased the cost of the alcohol threefold, pharmacists turned to the products of manufacturers, who could use more cost-effective methods. As the Civil War dragged on, more taxes, import duties, and licenses added to the burden carried by pharmacists. By the end of the war, an outraged Procter wrote that the large tax on alcohol (four dollars a gallon) "reaches nearly every important class of preparations made by the apothecary, and yields of itself a sufficient revenue to be drawn from this branch of business; but pharmacy has to carry, 1st, the duties on drugs, nearly all of which are imported; 2d, the right to carry on the business, by license; 3d, the right to sell brandy, whisky, and wine, for medicinal use, requires another license, although they are all officinal medicines; 4th, the stamp tax, which is a considerable item; 5th, the income tax; and now, lastly, a manufacturer's tax!" This last tax was the final irony. Because pharmacists generally sold stock quantities of preparations to local physicians for office or other use, they had to pay a fee for being a drug manufacturer.[86]

Not only did these taxes force pharmacists to use more factory-made drugs, but they also influenced other aspects of their practice. Because they had to obtain liquor licenses to carry medicinal alcohol, many pharmacists stocked beverage alcohol to pay for the license. Liquor brought in a clientele who wanted to buy tobacco and similar goods. Procter watched in dismay as the nature of apothecary shops turned more and more akin to general stores.[87]

Procter reacted to these and other changes that occurred during the 1860s by harking back to the progress American pharmacy had made during the 1840s and 1850s. To Procter, the key to this advancement lay in the control pharmacists had gained over the compounding of prescriptions. If every pharmacist would concentrate on improving his understanding of pharmaceutical art and science and apply this knowledge to practice, American pharmacy could again move forward.[88]

Other pharmaceutical leaders, however, had different ideas. Edward Parrish, for instance, thought that American pharmacists could best elevate their position through dignifying their dress and the appearance of their establishments. He suggested an ideal scheme whereby all the outward aspects of a shop, such as bulk windows and showcases, were to be eliminated and replaced by a

Procter's close friend and colleague Edward Parrish (1822–
72) was in partnership with his brother Dillwyn in a
successful retail and wholesale pharmacy in Philadelphia.
An outspoken advocate for pharmacy in America, Parrish
gained renown for his book *Introduction to Practical Pharmacy*. (From the American Institute of the History of
Pharmacy.)

suite of rooms. A patron would wait in a reception room while the pharmacist filled the prescription in an adjacent laboratory. Only medicinal and dietetic products would be offered for sale. In this type of practice a pharmacist would attract the level of respect that his vocation deserved.[89]

Neither Procter's nor Parrish's vision of the ideal pharmacy practice came to fruition. Pharmacy achieved its position in the framework of American medical care, not because each pharmacist became a medicinal chemist or eliminated all the trappings of commercialism, but through the gradual adoption of state-sanctioned paper credentials (state board certification). This new direction in the pursuit of professional status—group identity over individual achievement—came slowly to pharmacy during the remainder of the nineteenth century and into the twentieth, largely through the efforts of pharmaceutical organizations and from the requirements of law.[90] Physicians continued to dispense medicines, but wrote more prescriptions as the materia medica rapidly expanded. Crass commercialism also increased in an environment of free and open competition. And drug manufacturing passed into the hands of large companies.

For thirty years Procter watched the practice of pharmacy change under the pressure of unbridled competition, adverse economic conditions, war, new therapeutic trends, and the rise of large-scale manufacturing.[91] During this time he had practiced exceptional pharmacy, conscious of the ethical and scientific bases for his actions, as well as of his need to make a living. His views on important issues cannot be taken as typical for the period, but should be recognized as the leading edge of a movement to elevate pharmaceutical practice in the United States.

More than just a business, the small shop at Ninth and Lombard served as Procter's base of operations. From here he not only practiced pharmacy, but also performed scientific investigations, composed articles and lectures, edited the *Journal,* and carried out organizational work. By struggling and succeeding in his own pharmacy, Procter demonstrated to himself, his students, and his readers that one could practice in an ethical and scientific manner without financial ruin. In addition, his position as an everyday pharmacist gave wide influence to the views he expounded. But most of all, the parade of customers, physicians, students, and apprentices that went through his shop provided Procter with a continuous supply of new pharmaceutical and scientific challenges.

Pharmaceutical Scientist and Technologist

When Procter opened his pharmacy each day on Philadelphia's south side, he had more than just business on his mind. Leaving his senior apprentice or clerk in charge, he would retire to the rear of the shop to resume his scientific investigations. Joseph Remington recalled once going to Procter's shop for some advice. David Preston sent him to the back, where he found Procter apparently working on the process of percolation, "ensconced behind a small desk littered with papers and several bound volumes of the Journal of Pharmacy lying open in a convenient position for reference; it could easily be guessed that he had been surprised at his favorite occupation, and was writing an article for the Journal."[1]

Remington could have found Procter working on almost any subject that dealt with the art and science of pharmacy, including medicinal botany and chemistry. He was interested in every aspect of drugs: their natural history, collection, trade routes, cost, storage, and medicinal value. In addition, of course, Procter devoted himself to the study of how best to transform those drugs into elegant preparations for compounding into medicines. In essence, Procter was a consummate amateur scientist, curious about many things, and always ready to switch from one subject to another. Above all, however, he saw his investigations as an integral part of his efforts to advance the progress of American pharmacy.

This is perhaps the best of several artistic representations of Procter and his work. This painting by Dean Cornwell was based on the reminiscences of Joseph P. Remington and the fixtures in the museum of the Philadelphia College of Pharmacy and Science, which come from Procter's era. (Courtesy Wyeth, Inc.)

Because of the breadth of his interests, Procter's investigative work is not easy to characterize. The majority of his published papers dealt with small problems he met during everyday practice or those sent to him by readers of the *Journal*. When confronted with some difficulty in the compounding of a prescription, he would make note of it.[2] During a break in business, which occurred often in the early years at Ninth and Lombard, Procter investigated in a systematic way various methods for solving or avoiding a compounding problem. When he found a particularly novel solution, Procter wrote it up for the *Journal*.[3] Typically American in his attitude toward science, he followed a Baconian model of investigation, eschewing grandiose theorizing for what he called "the inductive path that leads to ultimate success."[4]

As a serious investigator, most of Procter's significant contributions arose from two major areas of work: the proximate analysis of plant drugs and application of the process of percolation to the problems of pharmaceutical preparations. In addition, out of his investigations of preparation methods came several important innovations in pharmaceutical technology.

For American pharmaceutical investigators of the mid-nineteenth century, proximate analysis offered the most attractive route to scientific achievement. In contrast to the restricted meaning of the term to chemists, pharmacists thought of proximate principles as "the several kinds of chemical constituents naturally formed and contained in plants and separable from the plant parts as well as from each other by the aid of different solvents." These principles were put together into groups, such as cellulose, gums, starches, pectin, sugars, albumin, fixed oils, organic acids, volatile oils, resins, and alkaloids.[5]

Procter's devotion to medicinal chemistry was not unique for his time, although highly unusual for an American. The work of Friedrich Sertürner, Pelletier, Caventou, and other European pharmacist-scientists of the early nineteenth century gave Procter and men like him an experimental model, which they used to break apart botanical drugs into components. To Procter and many of his contemporaries, most of the medicines in use in the mid-nineteenth century were "as unskilfully given as would be

the grinding of corn, cob, husk, stalk and root into meal, while
the kernel only can nurrish."[6] These early phytochemists sought
to obtain a pure chemical constituent, usually an alkaloid, which
could replace the crude drug in the preparation of medicines. In
contrast to the great variability of potency of a crude plant drug or
its preparation, the pure alkaloid offered physicians an agent of
reliable potency.

Sometimes an investigator could not extract the active principle,
but instead isolated it in a certain proximate fraction of the analyzed drug. He then tried to determine what outside circumstances
(maturity of the plant, growing conditions, and collection methods) and pharmaceutical techniques (selection of plant part, gar-
bling, and extraction methods) would yield the largest amount of
this fraction in active form. By sharing this information the pharmaceutical investigator could bring increased potency and uniformity to the materia medica. Such investigations benefited the
public's welfare through more dependable medicines and the pharmacist's pride by allowing him to continue the work of Sertürner
and his successors. Fame awaited the pharmacist who could isolate
a new alkaloid or proximate principle.[7]

A small amount of fame came to Procter upon publication of his
first scientific paper. For his student thesis, which the College
required for graduation, Procter followed the lead of previous
graduates and investigated a medicinal plant indigenous to North
America. Since before the founding of the College, prominent
Philadelphia pharmacists such as Elias Durand and Daniel B.
Smith had investigated the native flora. The rise of the botanico-
medical movement in the early nineteenth century further inspired
the small minority of scientific-minded pharmacists. Although
these pharmacists generally supported the efforts of the educated
elite of the medical profession to suppress rival sects, they shared a
common fascination with sectarians like the neo-Thomsonians
and the eclectics—the development of uses for North American
medicinal plants. Orthodox traditional doctors relied much more
on imported drugs. In addition to national pride, American pharmacists also promoted native botanicals, because they could be

obtained at lower prices and in far better quality than imported drugs.[8]

For his subject Procter chose *Lobelia inflata,* the main drug of the Thomsonian medical system.[9] That Procter would consider studying lobelia is indicative of the ambivalent attitude American pharmacists bore toward irregular medicine, as well as pharmacists' general disinterest in medical theorizing. For Procter and his colleagues, it was the *pharmaceutical* quality of a drug or preparation that mattered, that is, did it meet the standards for freshness and elegance of appearance. Whether or not a medicine had much therapeutic value was a secondary issue, that, even if not ignored by pharmacists, was still regarded by them as a *medical* issue. Conflicts among pharmacists concerning sectarian remedies did surface occasionally, but they were confined largely to debates about chemistry and nomenclature, rather than about therapeutic efficacy or theory. Moreover, for most pharmacists, Thomsonians, eclectics, natural healers, or other sectarians were simply customers and were treated no better or worse than anyone else coming in off the street.[10] Despite its controversial topic, the *Journal* published Procter's thesis in 1837 as it did other competent efforts.

As one disinterested in the sectarian battles within the medical profession, Procter was able to study the central drug of Thomsonian medicine objectively. From his tone Procter judged Thomson a quack, but his favorite drug, lobelia, had promise. Procter reasoned that "under the cognizance of the skillfull practitioner, it will be numbered among our most valuable remedies." Even if it were used by "empirics," the drug had profound physiological effects and deserved a closer look by regular physicians.[11]

This paper by Procter fit the standard form of a materia medica paper from the early nineteenth century. After an introduction, Procter described lobelia botanically ("Botanical History"), which included when to collect and how to store the plant. A section on "Properties" followed, describing its taste, gross physiological effects, its basic preparation method (for lobelia, it was infusion), and other physical characteristics and preparations. "Medical History" outlined lobelia's therapeutic properties as an emetic, diaphoretic, and expectorant. Procter discussed in brief the debate between the regulars and the Thomsonians, pointing out that the latter group was ignorant of lobelia's narcotic properties, but that regulars should not ignore lobelia's excellent usefulness as an emetic. He concluded this section with dosage information. (As an emetic, ten to twenty grains of the crude, one-half fluid ounce of the tincture, or a wineglassful of the infusion was administered.)[12]

The last substantial section forms the bulk of the paper, "Chemical History" or chemical analysis. Procter reported the results of fifteen experiments on lobelia, the tenth and eleventh yielding lobeline. (He called it "lobelina" following the naming convention of the day.) Lobeline had the common properties of alkaloids and exhibited outward characteristics similar to nicotine: "a light brown color, . . . the consistence of thick honey, a strong, somewhat aromatic odour, and a highly acrid, burning, nauseous, taste."[13] Overall the proximate analysis yielded the following constituents: gum, gallic acid, volatile oil, chlorophyle, "a green, fixed, oily, matter," a *"peculiar, alkaline, acrid principle,"* calcium and potassium salts, and lignin.

Procter applied a standard method for extracting an alkaloid in experiment ten:

Half a pound of the green plant was digested in one pint of water, acidulated with one drachm of concentrated acetic acid, for sixty hours, at a temperature of 70° Fahrenheit; after which it was subjected to decantation and expression.

This liquor, of which twelve fluid ounces were obtained, was saturated with pure magnesia, and filtered, when it was of a dark red brown hue, and had a very acrid taste. It was then treated with successive portions of sulphuric ether, at sixty Baume, until its acrimony was entirely removed.

The ether, after separation from the infusion, had a gelatinous consistence, owing, probably, to the presence of water, and some other matter.

To the ethereal liquor thus obtained, a small quantity of the chloride of calcium was added and the mixture agitated, to separate the water, and the ethereal solution obtained pure and colourless. This, upon evaporation, yielded a small portion of brown transparent matter, of the consistence of thick honey, having an intensely acrid taste, a strong, somewhat aromatic odour, with a decided alkaline reaction on reddened litmus paper.[14]

Procter drew several comparisons between lobeline and nicotine, including their gross chemical characteristics and yield from their plant sources. Moreover, he observed that the seeds contained twice as much lobeline as the leaf, a fact later utilized by those manufacturing the alkaloid.[15]

Procter had gone about as far as one could with his resources and equipment. In his first attempt at serious phytochemical work, he had achieved the brass ring—the isolation of an alkaloid. Yet his achievement failed to bring the usual accolades. Lobeline never became an important drug. Emetics were in decline as a therapeutic class, and lobelia was tainted by its association with the Thomsonians. Moreover, Procter did not have the equipment to do an ultimate analysis of the alkaloid, which would have given him clear priority for its isolation.

Still, after his initial work on lobelia, Procter pressed on with three more related papers during the next five years. First, he subjected the related species *Lobelia cardinalis* to proximate analysis. In this paper Procter reported isolating an alkaline substance, which he dubbed *lobeliana,* but it had no perceived medical utility. (In other words, its ingestion produced no outward physiological response.)[16] Second, he further investigated the nature of lobeline, paying special attention to the action of heat on the alkaloid and feeding the drug to a cat in order to compare it with nicotine. (Procter found lobeline far less toxic than the highly poisonous nicotine.)[17]

This third paper presented some new information, but appears mainly to have been a vehicle for Procter's claim of priority. Because lobeline never gained much use, however, the controversy over its isolation died quickly, with Procter accepted as the first to extract it.[18]

Procter concluded his investigations on lobelia with a study on various preparations of the plant drug.[19] Methods for making an extract, a vinegar, and a syrup of lobelia are described. He closed this paper, perhaps out of frustration, stating "It is believed that *Lobelia inflata* has yet to receive from professional men [i.e., physicians] that share of attention which it deserves and it is hoped by presenting to the practitioner the virtues of the plant in a condensed form, that he will take up the subject."[20] Despite his pleas, however, lobelia and its alkaloid remained a drug of interest only to the Thomsonians and some eclectic physicians.

During the early years of his scientific career, Procter aggressively pursued the analysis of other indigenous plant drugs. The plants investigated included *Xerophyllum setifolium* (turkey's beard), *Asarum canadense* (wild ginger), *Gaultheria procumbens* (wintergreen), *Betula lenta* (sweet birch), and *Monarda punctata* (horsemint). Procter's investigations into the nature of the last three plants mentioned were among the most successful and frustrating of his career. He found that with his meager resources of money and equipment, he could not bring his investigations to closure.

Gaultheria procumbens yields oil of wintergreen, one of the most popular flavoring agents. Procter had read of speculations that oil of wintergreen could be related to salicylic acid, so he tried his hand at a chemical investigation.[21] By reacting the oil with several reagents he demonstrated "that the oil . . . is a hydracid, forming salts with bases, and compounds with chlorine, bromine, and iodine, like saliculous acid; but at the same time, it exhibits differences in its reactions which render the identity of the two substances improbable." Procter concluded, "The only means of settling this question definitely is, to subject the oil and its compounds, to rigid ultimate analysis, which the want of accurate instruments has caused the author to defer to a future period."[22] (At that time, Procter was still working for Zollickoffer, saving his money to buy his own shop.)

Procter was correct. Oil of wintergreen is methyl salicylate. About a year and a half later, he reported that the Frenchman A. A. Cahours had done the ultimate analysis. Building on the work of Cahours, Procter demonstrated that the volatile oil of *Betula lenta* (sweet birch) also contains methyl salicylate. Furthermore, he studied the best ways to extract the principle from the bark of the sweet birch, laying the groundwork for a small industry. By the turn of the century almost all natural oil of wintergreen came from *Betula lenta*.[23]

Procter's early efforts in phytochemistry came to end in 1845 after he studied *Monarda punctata* and its volatile oil.[24] A few colleagues in both pharmacy and medicine had noticed that a camphor-like substance formed in containers of the horsemint oil. For purposes of discussion, Procter dubbed it *monardin*. After describing its physical properties (melting point, boiling point, etc.) and running some chemical tests on it, Procter concluded, "A careful, ultimate analysis of the oil, and of the deposit, would do much to throw light on this subject." A year later, A. E. Arppe identified "monardin" as thymol, getting his name in the reference works.[25]

Beginning in early 1846, Procter turned his focus away from proximate analysis toward tackling the problems of drug preparations and their standardization. Other laboratory interests came and went, but the study and application of the process of percolation, especially in the formulation of fluid extracts, continued until the end of his life. By establishing and popularizing this process, Procter sought to achieve several of his professional goals: to bring standardization to botanical drugs, to elevate the stature of pharmacists by improving the elegance of drug preparations, to keep the manufacture of drug preparations in the shops and out of large factories, and to ensure the quality of drug products and prescriptions for the public. Over time these goals varied in importance, but until his death Procter retained his faith that the full utilization of percolation would greatly enhance the progress of American pharmacy.

Although Americans fully developed this extraction technique, German and French pharmacists deserve credit for its invention.[26] The father-and-son team of Pierre and Polydore Boullay introduced basic nonpressure displacement (as percolation was first called) in 1833 before the Pharmaceutical Society of Paris. They demonstrated that when a measured amount of ground crude drug, dampened with a solvent (menstruum), was packed carefully into a percolator (a cylindrical container, in the shape of an elongated funnel), it would readily release its soluble constituents due to the hydrostatic pressure of additional menstruum being added to the column. Further improved by other Frenchmen, per-

In Elias Durand's shop on the southwest corner of Sixth and Chestnut in Philadelphia, Procter was introduced to the achievements of French pharmacy and the newest methods of pharmaceutical science. This shop also served as a meeting place for Philadelphians interested in medicinal botany. (From Joseph W. England, ed., *First Century of the Philadelphia College of Pharmacy, 1821–1921* [Philadelphia: Philadelphia College of Pharmacy, 1922], 101.)

colation came to the attention of American pharmacists in 1836 through the efforts of Elias Durand and his protégé, Augustine Duhamel.[27]

After composing the first article published in America on the new technique, Duhamel asked for and received the aid of Procter in tackling a more comprehensive study of the process.[28] By then (1839) both of them had been using percolation extensively, finding it quicker and more thorough than the traditional maceration technique.[29] In maceration, the crude drug was placed in a large bottle, covered with solvent, and allowed to soak, usually for several days. The pharmacist would then decant the resulting solution, press the damp crude drug to remove any remaining solution, and pass all the liquid through a filter. The whole process of maceration, pressing, and filtration took roughly as many days as percolation did hours. Duhamel and Procter contended that percolation worked better than maceration when preparing extracts of cinchona, rhubarb, digitalis, belladonna, chamomile, and gentian. Finally, the resulting preparations of the drugs were more elegant, that is, they possessed more of the sensible qualities of the crude drug (its taste and odor) without the murkiness generally associated with products of maceration.[30] Because of these advantages, Duhamel and Procter appealed successfully to the revisers of the *Pharmacopoeia* to "give this method their sanction" in the *USP* of 1840.

In 1842, after the *Pharmacopoeia* had recognized percolation as an optional pharmaceutical technique, Procter and Joseph Turnpenny wrote another article on percolation. Practical in approach, it suggested that pharmacists use percolation in preparing syrup of wild cherry, a particularly difficult preparation. Percolation simplified this process so significantly that the technique's popularity was virtually assured. Pharmacists across the nation began making this popular cough medicine this way.[31]

After turning away from proximate analysis in 1845, Procter aggressively championed percolation for the "rising generation of pharmaceutists."[32] By the 1850s many American pharmacists had adopted the new method, yet Procter had to defend it when it

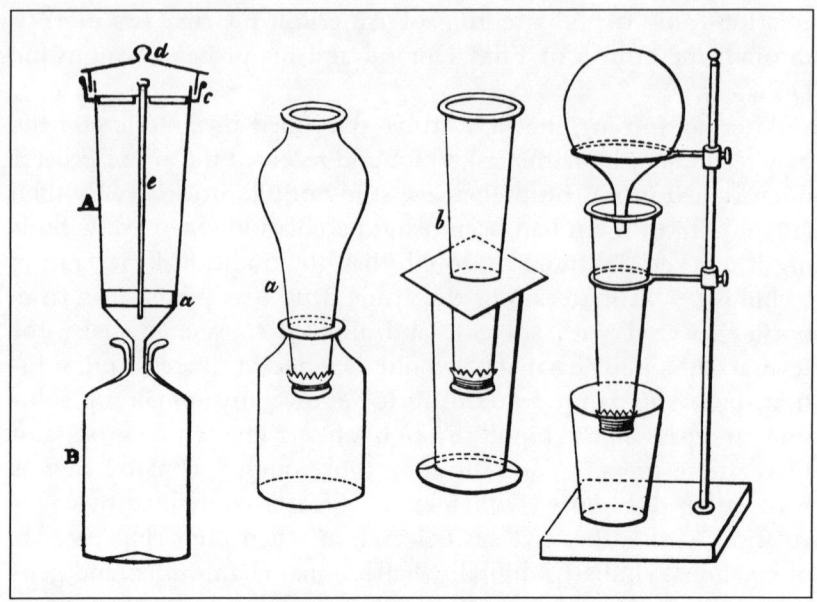

Procter was an early champion of the process of displacement (also called percolation). For those pharmacists who could not afford ready-made displacers like the one at the far left, he suggested using inverted lamp chimneys. The figure at the far right illustrates the set-up for continuous displacement. (From Francis Mohr, Theophilus Redwood, and William Procter, Jr., *Practical Pharmacy* [Philadelphia: Lea and Blanchard, 1849], 238.)

came under attack by those who questioned the ability of most pharmacists to handle the technique. As Edward R. Squibb pointed out:

The Pharmacopoeia very judiciously recommends that the process of percolation be restricted to the practice of those who may have had experience or instruction in manipulation. This restriction is rendered necessary by the circumstance that in the application of the process a very little bad management is liable to produce imperfect results, and be wasteful. The great convenience of the process had, however, led to its general adoption, regardless of the wholesome restriction, and it now only remains to be seen whether, by simple means, its general application may not be established without risk of bad results.[33]

An experienced pharmaceutical chemist, Squibb knew that if the drug was not ground to a uniform fineness and carefully packed into the percolator, the extraction process would fail. Inasmuch as most pharmacists in the 1850s did not possess high levels of training, formal or informal, they could not be expected to understand the principles behind the percolation process.

Once convinced of its applicability, American pharmacists used percolation to fulfill a variety of extraction needs. After Procter, along with a few colleagues, had worked out the standard method, he lost interest in complicating the process. He preferred to keep the official technique simple so that most pharmacists could do it without special equipment. Instead, he turned his attention to applying percolation to an emerging class of pharmaceutical preparations he all but created—the fluid extracts.[34]

In the 1850s fluid extracts became increasingly popular with physicians and pharmacists. George B. Wood, in the *United States Dispensatory,* referred to these extracts as "among the most efficacious, convenient, and elegant medicinal preparations."[35] What distinguished fluid extracts from other extracts was "the concentration of the active ingredients of medicinal substances into a small bulk, in the liquid form. Independently of the greater convenience of administration, the advantage of this class of preparations is that, the evaporation not being carried so far as in the ordinary extracts, the active principles are less liable to be injured by heat."[36] Procter had taken great care to develop formulas that were designed "to extract as far as possible all the valuable ingredients, and to condense them into the required bulk, of an ounce [of the botanical drug] to the fluid ounce [of extract] . . . in the way least calculated to injure their medicinal virtues and sensible qualities, leaving the resulting menstruum appropriate for retaining the active matter in solution."[37] This concentrated (and easy to remember proportion) made them especially handy for physicians to carry in their bags or cases.

Procter had been working on formulating fluid extracts since the late 1840s, with the purpose of keeping such manufacturing inside the pharmacy. With the subsequent rise of factory-made extracts, he concentrated on devising simple methods for preparing fluid extracts in the shop. If drug making could stay within the walls of the pharmacist's store, then quality (as well as the owner's reputation) would be kept high.[38]

In 1858 Procter accepted a query suggested at the annual meeting of the American Pharmaceutical Association that addressed the need for improved formulas for fluid extracts. Instead of answering in the usual two or three pages, he composed a twelve-page report that suggested detailed formulas for all the fluid extracts then popular in the Philadelphia area, totaling seventy-five in six different classes.[39] It is little wonder that in retrospect Joseph Remington went so far as to call Procter "the inventor of fluid extracts."[40]

Procter accomplished his goal when most of his suggested formulas were included in the *U.S. Pharmacopoeia* of 1860.[41] In recognition of this contribution to pharmacy, the association presented an award to Procter in the form of "an elegantly bound presentation copy of . . . Pareira's Materia Medica."[42]

Procter achieved a small victory for pharmacists by showing them how to make these convenient extracts with modest equipment. He purposefully had designed his procedures for small-scale, in-shop manufacturing.[43] Here is a typical formula:

Extractum Buchu Fluidum: Take of Buchu in fine powder, sixteen troyounces and Alcohol, a sufficient quantity. Moisten the powder with six fluidounces of Alcohol, pack it firmly in a cylindrical percolator, cover the surface with a disc of muslin, and pour in Alcohol until twelve fluidounces of tincture have slowly passed. Set this aside in a close vessel, and continue the percolation until two pints and a half more of tincture have been obtained. Evaporate this by means of a water-bath still, until reduced to four fluidounces, and mix it with the reserved tincture. After standing twenty-four hours, with occasional agitation, filter it.[44]

Procter's satisfaction, however, was short lived. The onset of the Civil War brought high alcohol taxes and stimulated the growth of large-scale pharmaceutical manufacturing. The pharmacopoeial

method for making fluid extracts (Procter's) required the evaporation and subsequent loss of large amounts of high-priced alcohol.[45] A drawn-out debate followed in the pharmaceutical press on
how to redesign extraction methods for the individual pharmacist.
In the end the *Pharmacopoeia* of 1870 adopted a compromise proce-
dure for making fluid extracts that required long maceration and
some savings of alcohol, but not enough to persuade an average
pharmacist to make his own extracts.[46] Large pharmaceutical
companies, even those which conscientiously followed the *Pharmacopoeia,* possessed considerable advantages over the corner
druggist: they could more efficiently grind drugs to the uniform
fineness required for optimum percolation; they were not under
the time constraints of prescription practice and could macerate
longer; and most important, they had the equipment needed to
distill off alcohol at low temperatures. While Procter debated in
the pages of the *Journal* with those who wanted to reform standard
methods for making fluid extracts in the shop, pharmacists were
setting aside their percolators in favor of ready-made extracts.[47]

Procter could see that the trend was moving away from his ideal
practice of pharmacy in which percolation played a significant
role. Reacting to the introduction of another line of fluid extracts
by a large-scale manufacturer, Procter reminded his readers:
"There is much to be learned in the relation of solvents to organic
matter in the process of percolation, and its practice is so entirely
adapted to the shop and within the ability of the pharmaceutist to
study with care and advantage, that it is greatly to be desired that it
will be retained as the process of solution *par excellance,* and not be
substituted by mechanical methods dependent on costly apparatus, and which necessarily throws the preparation of many important classes of medicines into the hands of large manufacturers."[48]

In the end, Procter lost this battle. Pharmacists turned more and
more to factory-made drug products throughout the nineteenth
century. The next generation of pharmaceutical leaders, like
Joseph Remington, emphasized the individual "tailoring" of each
prescription to the needs of each patient, but not the in-store
manufacture of ingredients.[49] The calling's pride in such ability
links back into Procter's efforts to make each pharmacist compe-

tent to prepare the ingredients for prescription compounding without overreliance on large pharmaceutical firms.[50]

In addition to his major efforts in proximate analysis and the application of percolation, Procter had some other continuing laboratory interests during his career. He was fascinated by iron and its preparations, both as a tonic and as an antidote. In 1838 Procter read of the success a French pharmacist had preparing ferrous carbonate of a uniform strength. Encouraged by the Frenchman's results, Procter tried his own hand at making a variety of iron salts used in pharmacy.[51] From the 1840s Procter made iron preparations a specialty, with some of his formulas becoming official in the *U.S. Pharmacopoeia*.[52] In that age of chlorosis, physicians often prescribed iron and its salts, usually in the latest form.[53] When Philadelphia physicians turned to French-style reduced iron in the 1850s, Procter became well known in Philadelphia for the high quality of his product. When he found that production of reduced iron absorbed more and more of his time, Procter turned over its manufacture to some colleagues.[54]

With arsenic commonly spread with butter on bread as rat bait, accidental poisoning abounded during the mid-nineteenth century, and iron oxides proved to be excellent antidotes. Procter saw, however, that old stock solutions of the oxides could be ineffective. In a series of experiments he demonstrated that while a nine-month-old solution of ferrous peroxide combined with a measured amount of arsenious acid in about four hours, a freshly made solution took only five to eight minutes.[55] By suggesting a new, faster way to make an iron antidote for arsenic and other heavy-metal poisoning, Procter may have saved several lives.[56] Although Procter's work on iron preparations did not attract the kind of attention obtained by the discovery of a new alkaloid, it probably accomplished as much good for society as any investigations he did.[57]

Better methods for making reduced iron are just one type of the many technological innovations Procter introduced to American pharmacy. In his 1849 edition of *Practical Pharmacy* (see chap. 6), Procter provided American pharmacists with ways to make their operations more sophisticated without much added expense. He

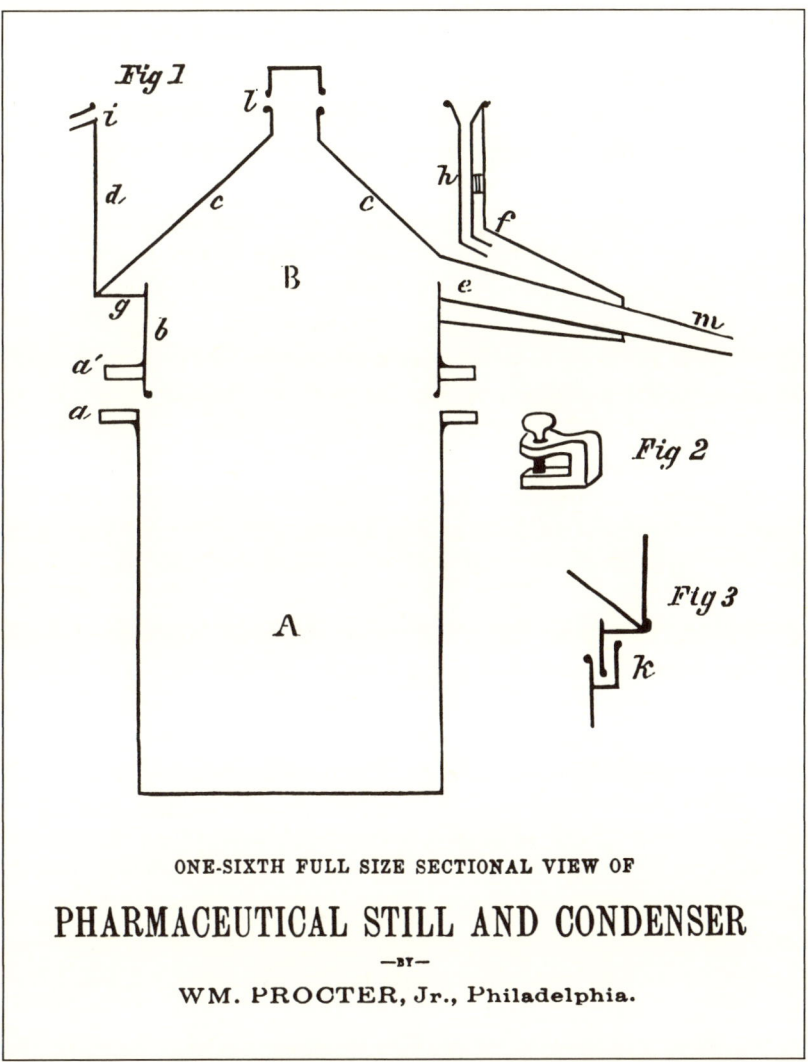

ONE-SIXTH FULL SIZE SECTIONAL VIEW OF

PHARMACEUTICAL STILL AND CONDENSER

—BY—

WM. PROCTER, Jr., Philadelphia.

In 1863 Procter answered a query asking for the best design and material for a shop still of two to four gallons' capacity. His design of a tinned iron still was the basic model used in the United States for the next thirty or more years. (From *Proceedings of the American Pharmaceutical Association* 11 [1863]: 209.)

did this by carefully describing in words and pictures those small improvements in pharmaceutical apparatus that an individual pharmacist could make with his own meager resources. Some of these included better blowpipes for analytical work and glass blowing, percolators made from extra lamp chimneys, inexpensive generators of hydrogen and carbon dioxide, and a laboratory still designed specifically for a cramped drug shop.[58] Although slightly modified with time, Procter's still remained the standard model for American pharmacies for over twenty-five years.[59]

Because he was a scientific dabbler as much as a serious investigator, Procter kept up with developments in other sciences. One of the innovations that he brought to the *Journal* was a new department called "Varieties," consisting of abstracts from scientific periodicals in several different fields. Procter found inspiration for some of his scientific investigations from this work. He reviewed scores of periodicals, including the *American Journal of Science, Medical World, Journal of the Franklin Institute, London Chemical Gazette, Medical and Surgical Reporter, British Medical Journal, Medical Times and Gazette, Repertoire de Pharmacie, Eclectic Medical Journal,* and the *Revue Pharmaceutique.*[60] If something in his reading piqued his interest or appeared questionable, Procter would take note of it and experiment on it later.[61] It did not have to be an earthshaking discovery. For instance, nearing his fiftieth birthday, Procter composed a three-page piece on the chemistry of soap bubbles and "pharaoh's serpents" (small, black disks that expanded into long black "snakes" when ignited).[62] He hoped that his own example of spontaneous experimentation might be followed by other pharmacists, contending that "the most trivial incidents may originate discoveries far out-reaching in their influence."[63]

As editor of the *Journal,* Procter pointed out the successes of other scientists and enthusiastically described how their inventions aided the progress of American society. His discussion of these topics reveals his broad working knowledge of the science and technology of his era.[64]

Despite his wide scientific interests, Procter kept his own serious work within the defined bounds of his calling. If pharmaceutical scientists wanted recognition for their speciality, they had to limit themselves to studies that involved, in some way, the preparation of medicines.[65] Procter insisted that pharmacists involved in investigations should publish their methods and results in full for the benefit of society, not hide their techniques for later personal gain. He chastised those who tried to reap "pecuniary benefit" from scientific discoveries. This stance was common among the leaders of America's scientific communities.[66]

What was uncommon was Procter's rejection of monetary gain from scientific or technological developments. A rather timid businessman, he failed to take advantage of any of his own inventions or formulas. Instead, Procter sought respect from his peers and the personal satisfaction he received from developing more elegant methods of medicine preparation. Perhaps this motivated his concern with the issue of priority, whether for his own discoveries or those of others. Scientific discovery was a matter of principle, not money, to Procter.[67]

Procter revealed some of his motivation for scientific work in his obituary of the Philadelphia chemist Robert Hare. Although respectful of his old friend, Procter criticized him for wasting time constructing complex apparatus, while others "achieved the immortality which had been waiting his acceptance."[68] As a man of "small means," Procter reminds us of Carl Scheele and others who had made discoveries without substantial support. His words also reveal an appreciation for the kind of "immortality" that a scientific investigator can seek.

Procter never acquired a reputation as great as Robert Hare, except within American pharmacy. From the publication of his first scientific work in the late 1830s to the death of his pupils in the early twentieth century, Procter was equated by pharmaceutical leaders with the best in research. His thoroughness and dedication to purpose became a model for a generation of pharmaceutical researchers.[69]

Yet, however strong Procter's yearnings for recognition were, his writings reveal more deep-seated motivations. He possessed a persistent curiosity about the natural world around him. A childhood companion recalled how young William would "interest other boys in stones that he would pick up in the streets, or in general subjects that would arrest his own mind."[70] Under the encouragement of his preceptor Zollickoffer and the pharmacist-investigator Elias Durand, Procter began his own investigations upon reaching maturity.[71] From the influence of these men, and through emulating others like Franklin Bache, Procter adopted an ethic of hard scientific labor. In an era when no one recognized pure relaxation as a virtue, experimentation also provided young Procter with a productive way to spend free time in his new shop.[72]

Science in mid-nineteenth-century America was largely an individualistic enterprise, which might explain some of its attraction to Procter, a shy man raised as a Friend. Quaker discipline encouraged an individual to dedicate oneself wholeheartedly to a practical cause. For some Friends the cause might be temperance or the abolition of slavery, but for Procter it was the progress of pharmacy. Pharmaceutical science and technology served as prime tools in that endeavor.[73]

The city of Philadelphia provided the perfect soil for Procter's fertile mind. By the mid-nineteenth century, Philadelphia had maintained its position of leadership in American medicine for several decades. The young field of dispensing pharmacy found an early home there. Although more drugs were imported through New York, Philadelphia wholesalers and manufacturers controlled more of the overall commerce in drugs and medicines. This city's intellectual opportunities attracted men like Elias Durand and Israel Grahame away from other Atlantic Coast cities.[74] In Philadelphia, Procter could attend lectures not only at the College of Pharmacy, but also at the Medical Department of the University of Pennsylvania.[75] His interaction with Zollickoffer, Durand, and lecturers such as George B. Wood, Franklin Bache, Robert Hare, and Joseph Carson opened to him diverse approaches to science and technology.[76]

The work of Philadelphia role models inspired several of Procter's early investigations. For example, in 1838 Procter read a paper by Robert Hare on the action of sulfuric acid on the volatile oils of sassafras, cinnamon, and cloves. Seeing value in Hare's technique,

Procter carried out similar experiments on twenty-eight oils, including camphor oil. He found that camphor oil contained little camphor and that the specific gravities of the tested oils provided "a good test of their purity and age."[77]

After Procter joined the Philadelphia College of Pharmacy as a member in 1840, he became involved in Philadelphia science in earnest. In 1841 he organized, with a group of colleagues, regular pharmaceutical discussion meetings held at the hall of the College. This group modeled their activity after the Society of Pharmacy of Paris.[78] During the 1840s participants in the meetings (members and graduates of the College) came to rely on Procter's expertise as a medicinal botanist and chemist. Procter's shop became a magnet of sorts for exotic drugs (see chap. 2), samples of which he brought to the pharmaceutical meetings. Occasionally, when disputes arose over the comparative value of related botanicals, Procter would be called upon to settle the issue in his laboratory.[79] Even though the pharmaceutical meetings declined during the 1850s and 1860s, they provided Procter with opportunities to hear of new research and with a sounding board for his own proposals.[80]

Procter also kept in touch with the wider scientific world, both in the United States and in Europe. In 1840, for instance, western explorer Josiah Gregg sent samples of Native American drugs to Philadelphia, where they eventually found their way to Procter's laboratory for analysis.[81] Twenty years later Procter did similar work for Philadelphia physician S. Weir Mitchell.[82] In 1853 Procter visited the Smithsonian Institution while in Washington and met its head, Joseph Henry, and his assistant, Spencer Baird. Procter later carried on correspondence with Baird.[83] From the founding of the American Pharmaceutical Association in 1852, the number of Procter's scientific contacts increased and, combined with his editorship of the *Journal,* helped make him a national figure in pharmacy. On the international scene, the medicinal botanist Daniel Hanbury, his counterpart in England, served as Procter's main contact.[84]

Though he enjoyed his international scientific contacts, Procter possessed a strong sense of nationalism, which helped to motivate his scientific investigations. In an era when Americanism reached a

peak, with cultural and scientific independence from Europe pressing on the minds of American intellectuals, Procter dedicated himself to the examination of indigenous drugs.[85] This uniquely
American field of pharmaceutical study corresponded with a
"widespread national ambition to make the nation entirely self-
supporting."[86] Throughout his career as editor of the *Journal*,
Procter championed both indigenous drugs and the cultivation of
foreign botanicals on American soil.[87] He also supported scientific
expeditions that might discover drugs anywhere in the New
World. He expressed some enthusiasm for the efforts of the eclectics to study the drugs of the American West, but criticized their
pharmaceutical methods as naïve and impractical.[88] Procter and
his contemporary pharmaceutical investigators were strong pa-
triots, but they did not openly voice resentment of European scientific superiority, as did Americans working in some other
fields.[89]

Hand in hand with the motivation of nationalism went his previously mentioned desire to see science aid the progress of both
pharmacy and society at large.[90] Science, especially chemistry,
provided the individual pharmacist with the knowledge necessary
to carry out his duties properly as the preparer of medicines.[91] For
this to be most successful, Procter urged students of pure science
to apply themselves to the processes of drug manufacturing in
order to "have a direct bearing in improving the facilities of life."[92]

Though emphasizing the practical, Procter encouraged more
theoretical work by opening the pages of the *Journal*. When some
readers chided him for publishing such papers, Procter responded:
"Let us reach down to the aid of those who can only appreciate
pharmaceutical literature in the shape of money-saving formulas
and processes, and give them a sustaining draught, mayhap, in
time, they may gain strength and interest enough to digest articles
of a higher grade, become pharmaceutists in reality, as well as in
name, and return the gift with interest in the form of papers to this
or some other Journal."[93]

Even though Procter distinguished theoretical from applied science, he saw both as critical to the progress of pharmacy. This
broad acceptance of all knowledge that might benefit pharmacy is

demonstrated by one of the innovations Procter brought to the American Pharmaceutical Association, the Report on the Progress of Pharmacy. In 1857 Procter chaired the association's Committee on the Progress of Pharmacy and composed its first report. In twenty-nine pages covering the topics of inorganic chemistry, organic chemistry, materia medica, and practical pharmacy, Procter abstracted what he discerned as the most important books and articles of the previous year. Although subsequent authors modified the format of the annual report, Procter's example of including almost any scientific material relevant to pharmacy in this department remained unchanged for over fifty years.[94] In how he edited the report Procter exhibited a mixture of amateur enthusiasm for science and solid training in shop manipulations.

Procter, of course, was not alone in his application of science to the problems confronting American pharmacists. Since the 1820s pharmacists like Charles Ellis, Daniel B. Smith, and Elias Durand had performed scientific investigations, usually emphasizing medicinal botany. During Procter's productive years, 1835 to 1870, a small cadre of American pharmacists tried their hands at pharmaceutical science. Most of these men were, like Procter, talented amateurs whose diverse interests caused them to investigate many different aspects of pharmacy. A few, Edward Squibb and John Maisch for instance, had experience directing large laboratories and depended mainly on their chemistry skills for their livelihood. (John Michael Maisch (1831–93) was born in Germany, coming to the United States in 1848. A skilled analytical chemist and medical botanist, he taught pharmacy at both the New York and Philadelphia colleges of pharmacy.)

For good reason, much of Procter's overall reputation rested on his scientific accomplishments. His output of scientific papers on pharmacy was unrivaled by other Americans during his lifetime. Among his contemporaries only John Maisch achieved a comparable level of national and international recognition. One interesting (if imperfect) measure of Procter's contribution is provided by the *Catalogue of Scientific Papers* (1880–1900) published by the Royal Society of London. This select catalogue includes 31 papers by Procter, 24 by Maisch, and 14 by Elias Durand. (For comparison, the catalogue lists 45 papers by Joseph Henry and 127 papers by Robert Hare.) While no towering figure of science, Procter stood tall within his own small, specialized field.[95]

When Procter's efforts are compared to those of the top echelon among his pharmaceutical contemporaries, the quality of Procter's scientific work does not appear exceptional. He was not particularly innovative. In studies of the percolation process, for example, Squibb and Grahame made the major breakthroughs.[96] In the field of medicinal botany, Procter's efforts did not match the quality of those of John Maisch.[97]

Yet for all his limitations, Procter was still the consummate master-pharmacist, able to do all parts of the apothecary's work with ease and assurance. More than any other American pharmaceutical scientist of his day, Procter successfully applied the lessons of botany, physics, and chemistry to the daily practice of an average pharmacist. Through science, Procter found a way to participate in the general stream of progress in American society and a higher level of meaning for his daily work. Through his publications and teaching, Procter gained a measure of renown. Moreover he provided an example of what one dedicated individual could accomplish in his backroom laboratory and thus helped set the stage for the full development of American pharmaceutical science that occurred in the late nineteenth century. And lastly, Procter combined his scientific prowess with a successful tenure as the nation's premier professor of the practice of pharmacy (chap. 6), thus laying the groundwork for the entry of the pharmaceutical disciplines into academe.[98]

The Pursuit of Drug Quality 4

To William Procter the pursuit of drug quality epitomized the challenges confronting American pharmacy. Without drugs of consistent quality and potency, all of the carefully compounded prescriptions and cleverly designed preparations would not serve the needs of the public.[1] Directly or indirectly, this quest touched almost every facet of his career. As a practitioner he was among the minority who tested the quality of drugs for use in his own shop; as a scientist he developed tests for others to use; as a teacher he taught his students how to catch the deceptions of the drug adulterater; as an editor he made his journal a clearinghouse for new standards; and as an organizer he used drug quality as a rallying point for the founding of the American Pharmaceutical Association.

Because physicians and the public saw them as the guarantors of drug quality, conscientious pharmacists questioned both the purity and the potency of their prescription ingredients.[2] Procter observed in 1846 that the "boldness with which the adulteration of medicines is carried on could hardly be believed, were it not for the occasional revelations which occur from time to time."[3]

Today we have a tendency to assume falsely that, because most nineteenth-century drugs were of dubious efficacy, impurities and adulterations were of minor consequence. Granted, the addition of sand to sponges or bullets to opium in order to increase their weight and value was relatively harmless. However, common practices such as mixing cheap digitalis leaves with those of belladonna or the substitution of salicin for quinine no doubt led to serious results.[4]

Procter and his contemporaries differentiated adulterated drugs produced in North America from those coming from abroad. Native, or "home adulteration" as it became known, was viewed by Americans as less malevolent than that from foreign shores.[5] Early in the nineteenth century, American drug grinders and wholesalers often added a little "powder of post" (sawdust) to stretch out their product, a practice apothecaries tolerated.[6] The practice became more widespread as pioneers headed West to open up the American plains, for these travelers could not return to claim pharmaceutical fraud. It may be that the tremendous intake of drugs such as calomel attributed to frontier pioneers was due partly to the low potency of the drugs available.[7]

Throughout the 1840s the situation worsened as Europeans dumped their substandard drugs on the American market.[8] One pharmacist became so discouraged that he lamented that it had "become almost as difficult to procure reliable medicines as it is impossible to discover the philosopher's stone."[9]

The problem of adulterated drugs was linked with uncertain potency of crude drugs, chemicals, and preparations available from wholesalers. Even when a package of a crude botanical drug seemed to be free of adulteration, the potency of the substance varied considerably depending on the subspecies of the plant, where it was grown, its level of maturity, and even the season of its collection. Essential botanicals in the materia medica such as opium, cinchona, belladonna, jalap, hyoscyamus, ipecacuanha, and digitalis varied enormously in quality and potency.[10] Procter occasionally examined samples of cinchona bark and opium that appeared pure but on analysis yielded no quinine or morphine.[11]

The uncertain quality of botanical drugs called into question the potency of preparations made by the pharmacist. For example, the strength of laudanum could vary from inert to highly poisonous depending on the morphine content of the opium.[12] Moreover,

Edward R. Squibb (1819–1900) was a close friend and colleague of William Procter. They worked together in the revision of the *United States Pharmacopoeia* and in plotting the course of the young American Pharmaceutical Association. Although trained in a pharmacy, Squibb went on to become a physician, chemist, and drug manufacturer. (From American Institute of the History of Pharmacy.)

when a pharmacist compounded a prescription using a ready-made preparation produced from crude drugs of unknown quality, he could only guess the potency of the resulting medicine that he dispensed. As a physician, Edward R. Squibb saw a connection between the low quality of drug preparations on the American market and the rise of medical skepticism. He advised that in "the failure of medicinal substances to fulfil the indications for their use, let the medical man more frequently call in question the character or quality of the agent than the sufficiency of the principles upon which the use was based."[13]

Procter agreed with Squibb and perceived the implications for American pharmacy. He saw clearly that unless the quality of drugs going into prescriptions was assured, pharmacists could not obtain the confidence of physicians and the respect of the public. If American pharmacy was to rise above the level of a common trade, this situation had to improve.[14] Procter's pursuit of drug quality took many forms. He made drug assay a large part of his pharmacy practice and dedicated much of his spare time to developing new tests.[15] As a pharmaceutical scientist, he praised "detecting adulterated and spurious drugs [as] an ample field for the exercise of . . . experimental talents."[16] Above all, Procter helped improve drug quality through his contributions toward revising the *Pharmacopoeia of the United States (USP)* and the *Dispensatory of the United States (USD).*[17]

One centuries-old approach to improving the quality of medicines was the creation and maintenance of a published set of standards, often called a pharmacopoeia. Usually composed by one or more members of a medical elite (sometimes with the input of pharmacists), pharmacopoeias were sanctioned by governmental units interested in ensuring the quality and potency of the drug supply. Generally speaking, a pharmacopoeia defined what ingredients would go into pharmaceutical preparations and how they would be prepared. In addition, it defined the nature (usually the source, such as plant part) of the crude ingredients themselves.

During Procter's lifetime, the *Pharmacopoeia of the United States of America* served as the main focus for efforts to improve the nation's drug supply. When first published in 1820, this book's main mission was to clarify communication between physicians and pharmacists by defining the names of drugs and giving recipes for a small number of their preparations. Although recognized as official by medical and pharmaceutical authorities, the *USP* had limited legal status until the twentieth century.

The first known call for a truly American book of drug standards or pharmacopoeia was made by medical reformer John Morgan in 1787. He proposed that a pharmacopoeia for Pennsylvania be composed by the College of Physicians of Philadelphia. After ratification of the United States Constitution, the College sent out a circular to 100 physicians with the mission of putting together a national work that would ensure that apothecaries and physicians agreed on the nature of drugs that went into medicines. The Pennsylvania effort failed to produce a book of standards, as did similar state efforts in South Carolina and Connecticut. In 1808 the Massachusetts Medical Society succeeded in publishing its own state pharmacopoeia. This compendium was later adopted by New Hampshire, and its existence stimulated the efforts of physicians in other states, primarily New York and Pennsylvania, to organize a national standard.[18]

In 1820 the first *Pharmacopoeia of the United States of America* was published under the authority of a convention that drew only about a dozen medical men. If not for the diligence of one physician, Lyman Spalding, it might not have appeared at all. When the time came ten years later for the first revision of the *Pharmacopoeia*, rival groups in New York and Philadelphia each composed a new edition.[19] Surprisingly, such a serious schism within the small group of physicians actively concerned with drug standards did not doom the progress of the movement. Two young physicians, both professors at the Philadelphia College of Pharmacy, George Bacon Wood and Franklin Bache, effectively saved the *Pharmacopoeia* by writing an excellent companion volume: the *United States Dispensatory*.[20]

Stability came to the revision process of the *Pharmacopoeia* with the appearance of the *United States Dispensatory* in 1833, because of the symbiotic relationship between the *USD* and the *USP*. For roughly half of the nineteenth century, the authors of the *Dispen-*

satory, Wood and Bache, directly influenced the revision of the *Pharmacopoeia* through their dominant role on the revision committees appointed by the pharmacopoeial conventions of 1840, 1850, 1860, and 1870. Moreover, the income derived from the sale of the more practical and popular *Dispensatory* subsidized the publication of the *Pharmacopoeia.*[21] For all their success the authors of the *Dispensatory* knew that they needed the *Pharmacopoeia* as the basis of their commentary. Largely because of this relationship and the dedication of the Philadelphians involved, the *Pharmacopoeia* survived.[22]

Procter's involvement in setting drug standards began in earnest in 1839. With the second revision of the *United States Pharmacopoeia* approaching, the College of Physicians of Philadelphia organized a committee, led by Franklin Bache and George B. Wood, to consider what changes in the compendium seemed most prudent. Soon after the start of their work, Wood and Bache sought the practical advice of a few pharmacists. As their primary consultants, they selected William Hodgson, Jr., who had received his pharmacy training in London at the respected chemist's shop of John Bell, and young William Procter. Thus Procter became one of the first two pharmacists involved directly in the pharmacopoeial revision process.[23]

The records from that committee in the George B. Wood Papers at the College of Physicians of Philadelphia show the large extent of Procter's contribution. Although eventually there were four other pharmaceutical consultants, Procter's reports form roughly two-thirds of the surviving records. For the committee, Procter did mainly chemical evaluations of preparations or suggested the best methods of producing chemical drugs. In his reports of December 1839, Procter related his findings on the making of hydrocyanic acid, sulphuret of antimony, water of potassa, and extract of Krameria. For the extract, Procter advocated the new technique of percolation.[24] Later when the Philadelphia group became the de facto revision committee of the *Pharmacopoeia,* they hired Procter as their primary consultant. He occasionally attended committee meetings to exhibit products of his work. For example, on June 8, 1840, he came to a revision committee meeting with a variety of

samples of oil of wine made to the specifications of different pharmacopoeias.[25]

Although Procter, Hodgson, and the other pharmacists provided a great deal of help, Wood decided that more input from
pharmacists was needed. Early in 1840 he sent a letter to Daniel B.
Smith, president of the Philadelphia College of Pharmacy, asking
for the cooperation of its members.[26] The College responded by
setting up its own committee of six members: William Fisher
(chair), Thomas H. Powers, Elias Durand, William Moore,
John C. Allen, and Charles Ellis.[27]

At its first meeting, the pharmacy committee saw a need for an
energetic secretary and appointed Procter.[28] In addition to Procter,
two other young men were asked to sit in: Ambrose Smith,
nephew of Daniel B. Smith, and Augustine Duhamel, talented
protégé of Durand.[29] The committee wanted to take full advantage of this opportunity to prove its scientific mettle and recruited
the College's brightest young minds.[30] Soon after their entry to
this committee, these young men exerted their presence, becoming active, if ex officio, members.[31]

Once on the pharmacy committee, Procter and Duhamel
strongly advocated the introduction of percolation into the *Pharmacopoeia* as an alternative method for making some preparations.
The committee agreed, setting up a subcommittee consisting of
the two young pharmacists, plus Fisher. Procter and Duhamel
presented a full report on the subject at the very next weekly
meeting.[32] In their final report to the College, the committee
commented that "perhaps the amendment of the most value [to
the *Pharmacopoeia*] . . . has been the introduction of the process of
Displacement . . . in the preparation of a large class of formulae."
The committee set up a subcommittee of Fisher, Duhamel, and
Procter to write a description of the process for possible inclusion
in the *Pharmacopoeia*. Up to that time no process had been detailed
in the *Pharmacopoeia*. Their recommendations were adopted by the
committee, which ordered that percolation take a prominent place
in its final report.[33]

With considerable speed and dedication, the College of Pharmacy committee went through the entire *Pharmacopoeia,* making

proposals for its revision.[34] All of their individual reports were handed over to a subcommittee of Durand, Ambrose Smith, and Procter to collate.[35] As secretary Procter did almost all of this work, taking a month to write out the 350-page report.[36]

The high quality of the final report of the College of Pharmacy committee and the importance of the questions that it raised so impressed the revision committee that they decided to go over the entire *Pharmacopoeia* once again.[37] The pharmacy committee recommended that 55 drugs be removed from the *USP*, the monographs of 82 others be amended, and 65 new drugs added to the materia medica list. In the preparations section, they advised that 55 formulas be removed from the *USP*, 159 amended, 113 introduced, and 94 be left unchanged.[38] The revision committee followed most of the recommendations for additions—perhaps because they realized that pharmacists knew which drugs were increasingly prescribed—but the committee ignored almost all their calls for dismissals. Only seven drugs were dropped from the materia medica list. It may be that the committee regarded it as presumptuous for pharmacists to recommend taking away drugs from the armamentarium of physicians.[39] In addition, the pharmacy committee recommended forty-six pages of purity tests for chemical drugs, to be included in the *Pharmacopoeia,* based on tests in the *Pharmacopoeia Londinensis.*[40] A comparison with the tests later included in the second revision shows that some of these suggestions were utilized by the revision committee.[41]

The experiences of 1839 and 1840 significantly influenced Procter and his subsequent career. Not only did he earn a reputation as a top-notch pharmaceutical chemist, but he also made contact with the elite of Philadelphia medical and pharmaceutical science.[42] He became a favorite of leaders of the College of Pharmacy, who came to view him as their own discovery.[43] At the young age of 22, Procter had seen his work bear direct results when the *Pharmacopoeia* adopted several of his recommendations for standards of purity and percolation as an accepted method of extraction. He had also taken part in the first significant cooperative project undertaken by medical and pharmaceutical groups in

North America, the second revision of the *United States Phar-macopoeia*.[44]

Procter continued to take part in the revision of national drug standards, relating this work to other signs of progress like steam power and the telegraph. He hoped that the *Pharmacopoeia* would serve as a unifying force for pharmacy in America, perhaps lead-ing to a decennial pharmaceutical convention like the medical con-vention organized for pharmacopoeial revision.[45]

Aside from the unifying effects of pharmacopoeial revision, Procter understood the practical benefits of standards. For in-stance, when a physician prescribed a nonstandard preparation of a drug, the pharmacist had to choose which one of several available to use. In the case of highly potent drugs like aconite or belladonna this practice probably produced many an anxious pharmacist and poorly treated patient. In Philadelphia in 1847 Procter had to con-tend with three popular, nonstandard preparations of aconite.[46]

As the revision of the *Pharmacopoeia* for 1850 approached, Proc-ter and his colleagues at the College of Pharmacy again organized for their input to the process. Procter returned as secretary, but this time the committee failed to produce a detailed report as in 1840.[47] As it said in its introduction: "They [the College revision commit-tee] have not found it necessary to render so elaborate a report as that of the committee of 1840; its chief bulk consists in addi-tions."[48] This was an understatement. The 1840 report consisted of about 350 carefully written pages, elegantly bound; the 1849 report only took up forty-seven pages and was sloppily put to-gether. Perhaps Procter was too busy with other things, or maybe the desire to demonstrate the scientific prowess of Philadelphia pharmacy had waned in 1849. Wood and Bache had promised organized pharmacy representation at the 1850 pharmacopoeial convention and membership in the powerful revision commit-tee.[49] This situation may have dissuaded the College committee from spending as much time as they had in 1839 and 1840.[50]

The College selected Procter in late 1849 as one of its three delegates to the Pharmacopoeial Convention.[51] Of the thirty dele-gates that attended the supposedly national convention held in

Washington on May 6, 1850, five represented local pharmacy societies and eleven delegates were pharmacists or physicians from
Philadelphia.[52] Aside from policy matters, the most important
decision made by the convention was the selection of the ten-man
revision committee, which would do the actual work. The men
were to meet in Philadelphia, where three would constitute a
quorum. This final decision effectively put the revision of the
national drug compendium in the hands of the four Philadelphians: George B. Wood, Franklin Bache, Joseph Carson, and
William Procter, Jr.[53]

Procter brought a significant innovation to the pharmacopoeial
revision process of 1850 when he insisted on the publishing of an
inexpensive edition. Previous editions of the *Pharmacopoeia* had
been printed rather elegantly, costing about $2.50. Procter con-
tended that the majority of physicians and pharmacists "know
nothing of our Pharmacopoeia except as they learn it through the
dispensatories." He wanted American pharmacists and physicians
to be thoroughly acquainted with the *Pharmacopoeia of the United
States* and to eschew those of other nations.[54] After considerable
lobbying and arm twisting, Procter got the revision committee to
publish a one-dollar version of *USP III.*[55]

To improve the drug market further, the New York College of
Pharmacy called together a convention of colleges of pharmacy in
1851 to draw up a set of strict standards for certain drugs. In 1848
Congress had passed the Drug Importation Act, setting up a system for checking the quality of drugs coming into American
ports.[56] The new drug inspectors enforced the law with some
difficulty, however, because few botanical drugs had quantitative
standards of strength and purity. (The tests of the *USP* were qualitative and generally applied to chemical drugs.) Procter and the
Philadelphia contingent at the convention successfully proposed
that representatives from the pharmacy colleges should reconvene
in 1852 in the Quaker City. The American Pharmaceutical Association arose from that gathering.[57]

With the recommendations of the convention and the support of
some colleges of medicine, the government drug inspectors denied
entrance to tons of crude drugs. M. J. Bailey, a drug examiner for

nine years at the port of New York, claimed to have turned back roughly 900,000 pounds of drugs during his tenure.[58] Yet all was not perfect. Every change of national administration brought on firings of inspectors, who were often replaced with political cronies. One such incident so infuriated Procter that he wrote,

From a cotemporary [*sic*] we learn that Dr. Peirce, Examiner of Drugs at the port of Boston and author of a work on the adulteration of drugs, etc., having, in the opinion of the Secretary of the Treasury, had sufficient opportunity to learn the business of inspecting drugs[,] has been graciously permitted to discontinue his service to the government, that Dr. Joseph H. Smith, of Dover, New Hampshire, may have a chance to acquire some knowledge of the art and mystery of drug-judging, before his friend, the President, ceases to be controller general. The Honorable Secretary is evidently afraid that his agents will become too acute at detecting adulterations, and injure the revenue by too strict an interpretation of the law, to avoid which, he brings into office physicians from the interior, where opportunity for making acquaintance with pharmacology and chemistry is necessarily limited. When the inspectorships of flour, whiskey, etc., are filled with persons ludicrously unfitted for their duties, we can smile and pass on; but when the purity of the drug market is to be periled, that political partizans may be rewarded, it would be criminal to remain quiet.[59]

Within a few years political cronyism had reduced the effectiveness of the 1848 law to the point where leaders of organized pharmacy abandoned hope of effective enforcement.[60]

With the apparent failure of the 1848 drug law to solve the problem of adulterated and poor-quality drugs, Procter provided concerned pharmacists with the latest information for the testing of the quality of the drugs they purchased and sold. As editor of the *American Journal of Pharmacy,* Procter published in full the reports of the Philadelphia College of Pharmacy's committee on drug assay.[61]

Procter argued that the old standard of "pure and sound" for crude drugs be replaced by a chemical reference standard wherever possible. For example, cinchona bark should be required to contain a certain percentage of quinine to be allowed into an American port.[62] While this was not an unusual idea for drugs like opium or cinchona, Procter insisted that this approach be extended to other imported drugs, like belladonna and hyoscyamus. When some customs officials tried to reject a shipment of benzoic acid because it had been made from horse's urine (via hippuric acid) rather than from natural benzoin, Procter objected. He contended that benzoic acid, if pure, was acceptable whatever the source. Chemical composition should be the final arbiter. Procter advocated this approach for the *USP,* but the consensus of the revision committee was that most pharmacists did not possess the skills necessary for chemical analysis. Instead they added a few easy chemical tests and relied mainly on organoleptic descriptions of sound drugs.[63]

During the 1850s Procter accomplished some of his most valuable work in pharmaceutical science and art. He continued to develop drug assays, but put more emphasis on better techniques for preparing drugs. As described in the previous chapter, he hoped that an expansion of the fluid extracts in the *Pharmacopoeia* would bring needed uniformity to botanical drug products.

When 1860 approached, Procter and other men concerned with drug standards in America turned again to the problem of revising the *United States Pharmacopoeia.* In Philadelphia the College of Pharmacy set up a local committee of revision, as they had for the last two revisions, and selected Procter and two other delegates for the Pharmacopoeial Convention.[64]

Thirty men attended the Pharmacopoeial Convention held in Washington in May 1860, ten representing local pharmaceutical societies. If the physicians connected with pharmacy, such as George B. Wood, Franklin Bache, E. R. Squibb, Robert Bridges, and Joseph Carson, are added to their number, the strong influence of pharmacy is demonstrated. The balance between the two occupations of the revision committee had begun to swing in the direction of pharmacy, a trend that continued until the generation of drug standards was essentially controlled by pharmacists.[65]

After adopting Procter's motion calling for the printing of another "low priced edition of the Pharmacopoeia for more general distribution among pharmaceutists," the convention turned to more controversial issues. Standards for weights and measures had become a problem because of the efforts in Great Britain to consolidate the three pharmacopoeias of London, Edinburgh, and Dublin into one compendium. British medicine still influenced American practice.[66] A British abandonment of apothecaries' weights in favor of avoirdupois weights or the metric system would have made the *USP* the sole national drug compendium adhering to the traditional system of weights and would have made the exchange of British and American preparations difficult, if not impossible.[67]

Procter supported switching from the traditional troy weights (apothecaries') to the avoirdupois system, as the British seemed to be doing with their new pharmacopoeia. Alfred B. Taylor, a Philadelphia pharmacist who made the study of metrology a serious hobby, advocated a new standard based on the grain. (The grain, about 65 milligrams, was the only unit of measurement shared by the troy and avoirdupois systems.) E. R. Squibb favored formulas written in parts by weight, which would have allowed a pharmacist to use weights of any system. Franklin Bache advised delay until the British made their decision, or else adoption of the French metric system. After a lively discussion, the convention dropped the issue, leaving it for the revision committee.[68]

As had been the practice in previous revisions of the *USP,* the revision committee held its meetings in Philadelphia, this time at the home of Franklin Bache, chairman of the committee.[69] For the next two and one-half years the committee met almost every week to discuss changes in national drug standards (roughly 120 meetings). Five men shared the work: Bache, Alfred B. Taylor, Procter, Squibb, and Joseph Carson. Bache as chairman and Taylor as secretary handled the administration of the committee, along with some revision work. The bulk of the revising was done by the other three men: Procter and Squibb covering chemistry and pharmacy with Taylor's help; Carson reevaluating the materia medica.[70] Committee member George B. Wood attended only the first two meetings, but did make an effort to communicate with the committee by mail while away in Europe.[71]

All the main participants lived in Philadelphia except Squibb, who traveled from Brooklyn down to Philadelphia and, at personal inconvenience, attended over half the meetings.[72] Whenever

Table I. Weights

American pharmacists of Procter's era had to contend with two major systems of weights: apothecaries' (or troy) and avoirdupois (the weights of general commerce). (Metric weights were not commonly used in pharmacy until the late nineteenth century.) Although the system shared common terms (pound, ounce, drachm, and grain), only the small unit, the grain, was actually shared by both. Prescriptions were written and compounded in apothecaries' weights. For manufacturing purposes in the shop, however, pharmacists usually used avoirdupois weights because apothecaries' weights rarely came in sizes above one pound. (Crude drugs in bulk were also bought and sold in avoirdupois.) Because official formulas were written in apothecaries' weights, it was imperative that pharmacists take care in converting when using the larger avoirdupois pounds.

Apothecaries' Weights

1 pound = 12 ounces = 96 drachms = 288 scruples = 5,760 grains
   1 ounce = 8 drachms = 24 scruples = 480 grains
      1 drachm = 3 scruples = 60 grains
         1 scruple = 20 grains

Avoirdupois Weights

1 pound = 16 ounces = 256 drachms = 7,000 grains
   1 ounce = 16 drachms = 437.5 grains
      1 drachm = 27.34 grains

1 apothecaries' pound = .823 avoirdupois pound
1 avoirdupois pound = 1.22 apothecaries' pound

(1 gram = approx. 15.4 grains)
(.0648 gram = 1 grain)

Source: H. C. Wood, Joseph P. Remington, and Samuel P. Sadtler, *Dispensatory of the United States of America*, 17th ed. (Philadelphia: J. B. Lippincott, 1894), 1807f.

he came to the meetings, Squibb exerted a powerful influence on the proceedings because of his expertise in large-scale drug manufacture and his strong support of the pharmacopoeia.[73]

Squibb argued that the *USP* should abandon the old apothecaries' weights and measures and switch to formulas based on parts by weight. He had found in his factory that weighing large amounts of liquid ingredients was both easier and more accurate than measuring them by volume. Procter and the rest of the revision committee rejected this method as being too abstract for the average pharmacist, instead choosing to keep a modified form of the traditional weight system.[74]

In contrast with the debate surrounding metrology, the rest of the revision committee agreed with Procter that percolation should become the extraction method of choice in the *Pharmacopoeia*.[75] When it came time to compose the official guidelines for the process, the committee handed over to Procter the various reports on the subject they had received from pharmaceutical and medical societies.[76] Two weeks later Procter presented a short essay on percolation for insertion into the *Pharmacopoeia*, which the committee adopted without change.[77]

The committee also turned to Procter to sort out the scores of proposed new fluid extracts. They recognized Procter's expertise by allowing him to compose the formulas for this whole class of important preparations.[78] In addition to writing the directions for preparing eighteen new fluid extracts, Procter rewrote the monographs for the seven that carried over from the 1850 *USP.* Procter also contributed several formulas for iron preparations, plasters, liquors, and solid extracts.[79]

The revision committee relied on Procter, because he actively practiced pharmacy, to make up preparations according to several proposed formulas for evaluation by outside consultants. In one instance Procter prepared different versions of collodion and sent them to Dr. Thomas Hewson of the Pennsylvania Hospital for trial there. Procter related Hewson's experiences to the revision committee, which added the physician's favorite collodion to the *USP.*[80]

The care that the revision committee took in its task seemed excessive to some practitioners of pharmacy and medicine. In 1862, with the *Pharmacopoeia* two years late and no signs of imminent publication, Edward Parrish chastised the committee for its apparent inaction.[81] Procter retorted that the British had started their revision process in 1857 and were still far from finishing.[82] Finally, on January 24, 1863, the committee instructed chairman

Bache "to place the manuscript in the hands of the publisher."[83] In contrast to previous revisions, this committee had systematically rewritten much of the *United States Pharmacopoeia*. Only one formula in the entire compendium remained the same from 1850 to 1860.[84] Procter hoped that his contributions would make the *Pharmacopoeia*, not the *Dispensatory*, the standard for drugs and medicines in the United States.[85]

The combination of several circumstances diminished the extent of Procter's influence on the quality of drugs in the 1860s. With the breakup of the Union, Southern pharmacists and physicians had no stake in the new *Pharmacopoeia*. The British switch from troy to avoirdupois weights confused matters further, since the *United States Dispensatory* usually provided recipes from both the United States and British pharmacopoeias for common preparations. Practitioners of medicine and pharmacy relying on the *Dispensatory* had to take special care not to combine parts of formulas when compounding medicines, as they could in the past.[86]

Above all, the concurrent rise of the large-scale manufacturing and the heavy tax on alcohol during the Civil War years hampered Procter in his efforts to further the stature of American pharmacy through pharmacopoeial reform. To Procter, only the individual pharmacist could ensure drug quality by making his own preparations from basic ingredients. In his view, pharmacy was "the art of preparing medicines," not merely their dispensing.[87] He succeeded in popularizing fluid extracts, the focus of his efforts toward uniformity in botanical drugs, but failed to get most pharmacists to make their own.[88] The formulas for fluid extracts in the *Pharmacopoeia*, written by Procter, not only called for the evaporation of large amounts of costly alcohol, but also required "a degree of skill, a perfection of manipulation, and an honesty of purpose . . . none too common."[89] These circumstances caused pharmacists to buy fluid extracts and other preparations from large-scale manufacturers, who generally ignored the *Pharmacopoeia*. Thus the very preparations that Procter hoped would bring uniformity to galenical pharmacy actually increased the variability of medicines compounded by pharmacists.[90]

When Procter saw that pharmacists were using factory-made

fluid extracts as shortcuts in medicine making, he insisted that the revised *Pharmacopoeia* of 1870 contain formulas "marked for the simplicity and directness of their manipulation." Procter hoped that pharmacists would return to making their own extracts according to the *Pharmacopoeia,* thereby improving the predictability of drug therapy through more uniform medicines.[91]

Procter rejected the apparently inevitable trend of pharmacists' buying more and more official preparations from wholesalers. He thought that better organization and education would increase the vocational pride of pharmacists and inspire them to make their own medicines "from scratch." Along with other pharmacopoeial leaders, he accepted the cost-effective production methods of the industrial revolution only when applied to making chemicals or especially difficult preparations such as blue mass.

Procter failed to realize that the traditional approach taken by pharmacopoeial revisers of the nineteenth century had become outmoded despite the acceptance that the *Pharmacopoeia* had gained through the *Dispensatory.*[92] Oscar Oldberg described well the situation:

It is a rule, generally adopted by compilers and revisers of pharmacopoeias and pharmaceutical formularies, to select, for officinal works, such formulae and processes as will be most practicable to the pharmacist proper. Pharmacopoeias are not written for wholesale manufacturers. For such preparations as can be more profitably prepared on a large scale, no officinal formulae are, therefore, given in many pharmacopoeias.[93]

After 1865 pharmacists turned more and more to wholesalers for their prescription ingredients and less to their own laboratories. And aside from Squibb, no one involved in the pharmacopoeial revision process had direct connections with the rapidly rising pharmaceutical industry. Antagonism between the small-scale practitioners who wrote the standards and the large-scale manufacturers who made most drug preparations reduced the effectiveness of the *Pharmacopoeia* during the 1860s and 1870s.[94]

In contrast to the mixed results Procter saw coming out of pharmacopoeial reform, his contributions to the *Dispensatory* yielded more direct gains. He had long recognized the popularity of the *Dispensatory,* once describing it as

the text book—the codex—of the United States apothecary, more truly than the Pharmacopoeia; because, embodying, as it does, that work, very many apothecaries and physicians know the Pharmacopoeia only through the medium of the Dispensatory—a fact which, however honorable to the authors of the latter, is to be regretted, as depriving the former of that universally authoritative position which it should hold. On the other hand, it may be said that had it not been for the deep interest taken in upholding the authority and dignity of the Pharmacopoeia in their excellent Commentary, by the authors of the Dispensatory, that work would have been far less *influential in assimilating the practice of pharmaceutists* than it now is.[95]

To Procter, the centerpiece of drug standardization was this assimilation. If each pharmacist prepared official medicines in the same way, even if not the best way, the quality and potency of drug products became more dependable. Physicians could dose their patients more precisely and accidental poisonings would decline. Both the welfare of the public and the reputation of American pharmacists would benefit. During much of his career Procter hoped that the *Pharmacopoeia* would be the major force in this effort; he came to realize that its commentary, the *Dispensatory,* reached and influenced a far wider audience.

Since the late 1830s, Procter had done analysis work for George B. Wood.[96] As Procter began to contribute to the literature of pharmaceutical chemistry, Wood and Bache came to rely on his expertise. By the thirteenth edition of the *Dispensatory,* Procter was perhaps the most cited authority in the work, with his name appearing in over 130 different monographs. After the death of Franklin Bache in 1864, Wood asked Procter to update the pharmaceutical section of the book originally composed by Daniel B. Smith.[97] The forty-page outline of pharmaceutical manipulations Procter rewrote included some illustrations and text from his 1849 book. Through these pages in the *Dispensatory* Procter explained proper pharmaceutical technique to thousands of American physicians and pharmacists, many more than he reached with *Practical Pharmacy* or the *American Journal of Pharmacy.*[98]

Although largely composed by the same people for thirty years—Wood and Bache—the *Dispensatory* and the *Pharmacopoeia* were far different volumes. Standing side by side the *Dispensatory* dwarfs the *Pharmacopoeia*. For example, *USP IV* (published in 1862) is a small, manual-sized book of 400 pages. The print is large with plenty of white space. *USD XII,* the issue following *USP IV,* is a standard-sized reference work of about 1,700 pages of small type. Considering that the *USD* had about two and one-half times as many words per page, the *Dispensatory* held about ten times the information of the *Pharmacopoeia*. The descriptions of the materia medica (drugs) in the *USP* were exceedingly brief. For example: "Cinnamomum. *Cinnamon.* The prepared bark of Cinnamomum zeylanicum (Nees, *Laurin.*), and Cinnamomum aromaticum (Nees, *ibid.*)."[99] The *USD* had five full pages on Cinnamomum, including its various synonyms, its complete botanical description (both varieties), its "culture, collection, commerce, etc.," its medicinal properties, and a list of its officinal preparations in the United States and British pharmacopoeias.[100] An even more extreme example is opium. *USP IV* describes it in forty-two words, the *USD* spends twenty-four pages on this central drug of the era. (For preparations, the *USP* gave the formula and basic technique for manufacture, which the *USD* repeated, adding the recipe from the British pharmacopoeia, plus commentary on its properties.) Lastly, the *USD* was much more than a commentary on the *USP.* The *USD* of 1865, for example, contained almost 200 pages of "Drugs and Medicines Not Officinal" in either the United States or British pharmacopoeias, as well as a multitude of useful scientific tables and other data not contained in the *Pharmacopoeia*.

By the late 1860s Procter altered his approach to the pursuit of quality drugs. Instead of insisting that the *Pharmacopoeia* supersede the *Dispensatory,* he stressed their interchangeability.[101] He may have received positive reports of the effect his contribution to the *Dispensatory* was having on pharmaceutical practice, or perhaps he felt obligated to Lippincott, the publisher of the *Dispensatory,* which had probably paid him for his time. Moreover, his trip to Europe in 1867 had shown Procter how strong regulation could reduce the extent of poor pharmaceutical practices. Intraprofes-

sional efforts such as a gentleman's agreement to conform to the
USP seemed paltry when compared to laws on the Continent
calling for the inspection of pharmacies by authorities empowered
by the state. For whatever reason, Procter chose to take a smaller
part in revising the *Pharmacopoeia* of 1870. He attended the national
convention in Washington, but did not actively work with the
revision committee, with John Maisch taking his position on the
committee.

Reflecting on the work he and others had done in pursuit of
quality drugs over the previous thirty years, Procter concluded
that it had been a war worth waging, even if most battles had been
lost. The quality of imported drugs was as low as ever because the
Drug Law of 1848 was a dead letter, and the adulteration of medicines continued unabated. The *Pharmacopoeia,* once the centerpiece
of his efforts, had gained respect but not widespread use among
pharmacists and physicians. Ever positive in his writing, Procter
tried to encourage others to continue the pursuit of quality drugs
through pharmaceutical science (drug assay and potency standardization) and institutional activities (pharmacopoeial revision).[102]

Procter could not know that the revision following his death
(1880) would see a revival of some of his old ideals and a large step
taken toward their realization. With organized medicine losing
interest in the *Pharmacopoeia,* organized pharmacy rose to the occasion, virtually taking over the revision process. A scholarly hospital pharmacist from New York, Charles Rice, came to dominate
the revision process, putting it on a modern footing. The new
Pharmacopoeia contained the tests, assays, and chemical formulas
that Procter had envisioned in his ideal compendium.[103] Yet this
shift toward a pharmacopoeia thoroughly based on medicinal
chemistry was achieved by largely abandoning the traditional concept that the pharmacist should manufacture most of his preparations. Procter's dream that each corner druggist would become a
pharmaceutical chemist was not to be.[104] The fight that Procter
had led against the onslaught of factory-made preparations died
out gradually.

As a pharmacopoeial innovator Procter probably ranks behind
Edward R. Squibb, who accepted much earlier the future role of

the *Pharmacopoeia* as a guide to industrial production of medicines. Procter, however, played a crucial role in getting pharmacists involved in the pharmacopoeial process. The successes Rice and Remington, Procter's protégé, had in modernizing the *United States Pharmacopoeia* can be traced back to the efforts of Procter and his Philadelphia colleagues in 1839 and 1840. Moreover, percolation and fluid extracts continued to rise in popularity after Procter's death and did eventually lead to greater uniformity of botanical preparations.[105] Although it is difficult, if not impossible to gauge accurately how much Procter helped to improve the actual quality of drugs during the mid-nineteenth century, he carried the banner of drug standards as high as any of his contemporaries. He kept the dream alive until the next generation of pharmacists, whom he had helped educate and inspire, could bring it into being. Congress made the *Pharmacopoeia of the United States* and the *National Formulary* of the American Pharmaceutical Association official standards in enacting the 1906 Federal Food and Drugs Act. Both compendia committees were headed by former students of Procter, Joseph P. Remington and C. Lewis Diehl.

5

Editor of the
American Journal of Pharmacy

It can be debated which aspect of Procter's career most influenced American pharmacy: his key role in the founding of the American Pharmaceutical Association (APhA), his precedent-setting tenure as professor of pharmacy, his diligent efforts for improved drug quality, or his investigations in theoretical and practical pharmaceutics. All these achievements depended in part on the vehicle of the *American Journal of Pharmacy;* in 1851 and 1852, when the APhA took form, the *Journal* was the only national forum for the discussion of organizational issues. In his editorials, Procter all but set the agenda for its first meeting. As a teacher, Procter used the *Journal* to advertise for his College and sing its praises. And above all, the *Journal* published hundreds of Procter's papers dealing with pharmaceutical science and art.

In contrast to his role in the above developments, Procter played no part in the beginning of the *Journal.* It was already a stable pharmaceutical periodical, the only one in the United States, when he took over the editorial duties in 1850. Although he changed outward appearances of the *Journal* in only a few small ways during his twenty-year tenure, Procter fine-tuned and narrowed its focus, clarifying its direction. He made it the leading periodical in the United States for those interested in the pharmaceutical sciences. His even-tempered competence and dedication as editor won him the admiration of American pharmacists.[1]

As editor, Procter controlled one of the leading scientific periodicals in America.[2] Born as the *Journal of the Philadelphia College of Pharmacy* in December 1825, the *Journal* was christened the *American Journal of Pharmacy* in 1835 and thereafter published as a quarterly. Based on the model of the *Bulletin de Pharmacie* of the Society of Pharmacy of Paris, the *Journal* was the first of its type in English.[3] The *Journal* followed examples set by other American scientific journals like Silliman's *American Journal of Science and Arts* (f. 1818) and especially the medical and scientific journals based in Philadelphia: the *American Journal of the Medical Sciences* (f. 1818) and the *Journal of the Franklin Institute* (f. 1824). Like other early scientific periodicals in America, the *American Journal of Pharmacy* relied heavily upon active editors for vitality. From its connections with a resolute organization, the Philadelphia College of Pharmacy, flowed editorial talent and financial support. The energetic early editors set an important precedent for Procter; he would have to rely heavily on other periodicals and his own pen for copy.[4]

The content of the *Journal* during the 1830s and 1840s had reflected concerns of the editors, who were all physicians, but still kept close to interests of aspiring pharmacists: chemistry, botany, pharmaceutical manipulations, and to a lesser extent, natural history. The *Journal* reached only a small minority of America's pharmacy practitioners. The Philadelphia College of Pharmacy itself, both before and during Procter's life, consisted of a small group of idealistic pharmaceutists striving for vocational and scientific achievement in a city pervaded by quackery.[5] Never forming a majority of the drug dispensers in Philadelphia, this society received its strength and direction from a cadre of fifty members or so, a small number when compared with the hundreds of drug shops that existed in the Philadelphia area in the mid-nineteenth century. Through the *Journal* these men broadcast their message that by means of self-improvement American pharmacists could elevate their practice and gain some of the prestige and position accorded to their continental counterparts.[6]

When William Procter began contributing papers to the *Journal* in the late 1830s, its basic tenor and format were already in place. The editors had sought to provide its small readership with an up-to-date notice of the progress of pharmaceutical science and a medium for the publication of their own investigations. The success of the *Journal* came from the hard work of its editors, men like Joseph Carson.

Carson influenced Procter's early development as a teacher and editor. He taught Procter materia medica and pharmacy at the College's school of pharmacy, serving as a role model for Procter's own teaching later on. He continued as a co-worker at the school with Procter until 1850, when he accepted the chair of materia medica at the Medical Department of the University of Pennsylvania. At that time Carson handed over control of the *Journal* to his young coeditor. Procter's abilities probably showed themselves as a student under Carson at the school and were demonstrated by his contributions to the *Journal* and to the *U.S. Pharmacopoeia* in the late 1830s and early 1840s.[7]

Procter's first contribution to the *Journal* (1837) outlined his research on *Lobelia inflata* (discussed in chap. 3). With future reports on lobelia, iron preparations, and the constituents of wild cherry bark, Procter rose rapidly as a prime contributor to the *Journal*. With his growing importance in Philadelphia pharmacy came appointment to the publishing committee of the *Journal,* elected from the membership of the College.[8] Several of the most prominent figures of Philadelphia pharmacy served on this committee, including Charles Ellis, George B. Wood, Franklin Bache, Augustine Duhamel, Robert Bridges, Ambrose Smith, and Elias Durand.[9] Procter was 24 years old.

Procter's informal editorial involvement increased gradually after 1841. For example, Procter reviewed the new, American edition of Graham's *Elements of Chemistry* in 1843.[10] In addition to writing reviews, Procter did some editing before 1848. For instance, he added editorial comments in a footnote to an article on extract of Indian hemp.[11] Such activities were stimulated by Procter's energetic participation on the publishing committee. In an editorial announcing the appointment of Procter as coeditor, Car-

Joseph Carson (1808–77) taught Procter materia medica and pharmacy at the Philadelphia College of Pharmacy in the late 1830s and introduced him to pharmaceutical journalism in the 1840s. Carson began his career as an apprentice apothecary and later graduated from the Medical Department of the University of Pennsylvania. (From Joseph W. England, ed., *First Century of the Philadelphia College of Pharmacy, 1821–1921* [Philadelphia: Philadelphia College of Pharmacy, 1922], 117.)

son implied that his new assistant would be taking on some of the duties of the publishing committee. That committee, however, was to "continue its supervision" of the *Journal*.[12]

There exists no published account of why Carson chose Procter as his coeditor, but it it not unlikely that the paths of the *Journal* and of Procter converged during the 1840s until they united. In that decade Procter became the *Journal*'s most prolific contributor, in terms of both original and review articles. Apart from its sheer volume, this contribution had both quality and breadth of scope. Procter enjoyed the theoretical and the practical, the scholarly and the popular. While previous major contributors had stuck to their specialities—Bache and Bridges on chemistry or Wood and Carson on materia medica—Procter wrote on both these and on strictly pharmaceutical subjects. In addition, the *Journal* work may have lessened the disappointment Procter experienced with his young, struggling business.

With his election to the newly created chair in practical pharmacy at the College's school of pharmacy in 1846, Procter took over part of his old teacher's duties. Carson, a trained apothecary as well as physician, gladly handed pharmacy over to Procter and concentrated on his first love, materia medica.[13] Even though it is difficult to measure the importance of their interaction on the College's faculty, the level of Procter's contributions to the *Journal* increased in volume and scope after 1846; his name appeared in larger type on the *Journal*'s title page (with Prof. Robert Bridges), and he began extensive editing chores soon after.

Procter as Editor

In 1848 the first "editorial department" appeared in the *American Journal of Pharmacy,* which announced the appointment of the new coeditor, William Procter, Jr. It is fair to suppose that Procter initiated this new department. For the next two years most of the editorials seem to have come from his pen rather than from Carson's.[14] As discussed above, Procter followed the progress of British pharmacy closely. It is likely that Procter convinced Carson to include this new department based on the example of Jacob Bell's editorials in the *Pharmaceutical Journal*. Procter's first change in the *Journal's* format, the addition of editorials, proved the most significant.[15] It gave him the platform needed to advocate his program of progress for American pharmacy.

The coeditorship of the *Journal* may have continued indefinitely if George B. Wood had retained the chair of materia medica at the Medical Department of the University of Pennsylvania. The department elected Carson to succeed Wood and he accepted, serving from 1850 to 1876. Procter, an obvious choice to replace him as editor, was elected unanimously by the College Board of Trustees.[16] For the next two decades the *American Journal of Pharmacy* provided a framework for Procter's life. Its deadlines marked his calendar, its empty forms beckoned for copy, its creditors demanded money, its subscribers sought satisfaction. Like his Philadelphia editorial colleague Isaac Hays, Procter possessed a quiet, yet firm character that carried him through difficult times. For Procter the *Journal* provided many things: a little extra income, which helped in the early 1850s when his business lagged; a position of recognition within the scientific establishment of Philadelphia; and a ready outlet for his scientific papers and personal opinions on pharmacy. During the 1850s and early 1860s, before his shop became busy and his health declined, the *Journal* gave Procter an added sense of worth and accomplishment.[17]

In 1850 Procter assumed control of a relatively stable publishing enterprise. The format of the journal had been set early in its history and had changed little since the mid-1830s: each quarterly issue consisted of original contributions, reprinted articles (largely foreign), pharmaceutical notices (collections of hints and tips on shop manipulations), minutes of College meetings, and abstracts from medical and pharmaceutical journals (again substantially foreign). The original contributions rarely discussed scientific theory, but usually dealt with concrete observations or reproducible laboratory experiments. This characteristic prevented these articles from becoming foci for the heated arguments and retaliatory hyperbole common in American medical journals of the 1830s and 1840s.[18] As Procter might have put it, the *Journal* was dedicated to the progress, not the revolution, of pharmacy.

As the first pharmacist-editor of the *Journal* since its beginnings under Daniel B. Smith (1825–28), Procter pledged "to preserve its scientific standing undiminished, and to increase its practical usefulness."[19] Procter accomplished this goal chiefly through the addition of two new departments. The first, called "Varieties," appeared in the first issue edited solely by Procter. Similar to the old "Miscellany" department it replaced, "Varieties" usually contained five to ten pages of short abstracts taken from scientific periodicals. Unlike "Miscellany," "Varieties" often included long extracts from recent books. From 1854 through 1856, for instance, about half of each "Varieties" section reprinted a translation of *On Perfumery* by Septimus Piesse.[20] With this department Procter kept pharmacists aware of the spectrum of mid-nineteenth-century science, from electric illumination to the density of ice, the strength of cotton fibers, the use of mercurial ointment as an anthelmintic, and the acclimatization of ostriches in the Alps.[21]

To fulfill his wish to make the *Journal* more practical, Procter began a department called "Gleanings."[22] Just as "Varieties" had generally avoided purely pharmaceutical subjects, the "Gleanings" stuck close to practical pharmacy. In these abstracts from chemistry and pharmacy journals, Procter emphasized new apparatus, preparations, and assaying methods. Like the abstracts in "Varieties," these "Gleanings" were kept short; but unlike the older department, "Gleanings" had a more personal flavor. For example, if particularly interested, Procter would try out the reported new dosage form, extraction process, or assay and then comment.

In a way, "Varieties" and "Gleanings" represent well the two ends of Procter's spectrum of interest within pharmacy. Educating pharmacists about the progress of science (and their potential part in it) through "Varieties" was a labor of love; "Gleanings" reflected Procter's lifelong affection for pharmaceutical "hints and tricks."

Procter relied on the British *Pharmaceutical Journal* for material for both abstract departments, but he also searched scores of medical and scientific periodicals, both foreign and American. Competent in French, Procter extracted a great deal from the excellent French journals of the day, especially the *Journal de Chimie Medical* and the *Journal de Pharmacie,* which were apparently on the exchange list of the *American Journal of Pharmacy.* (These journals occasionally contained abstracts of articles from the *American Journal of Pharmacy.*)

During his tenure as editor of the *Journal,* Procter brought about other changes. He expanded it from a quarterly to a bimonthly in 1853.[23] During this same period he further increased the amount of copy in the *Journal* by reducing white space and type size and by simplifying the index.[24] The combination of these changes nearly doubled the volume of text appearing in the 1854 volume compared to that of 1850. In 1852 the *Journal* added advertising pages, containing notices from Philadelphia wholesalers and equipment manufacturers. (Procter exercised tight control on the advertising, with no advertisements for nostrums appearing.) Procter's innovations paid off with an increase in subscribers and the approval of the publishing committee.[25]

Procter's efforts to get more practical contributions into the *Journal,* however, fell short. He had hoped to make the *Journal* a forum for practical debates within the context of pharmacy practice. He suggested starting a new department that would be devoted to letters from readers "relative to the state of pharmacy—notices of adulterations—unusual phenomena in compounding prescriptions—incompatible prescriptions—the changes that officinal preparations undergo in warm latitudes—new remedies, etc." No such department ever arose, indicating a low response to Procter's call. On rare occasions letters of this type did appear on Procter's desk, and he published them, often with his own response.[26]

Procter had more perplexing problems. As editor, he also had to supervise the business side of the *Journal,* which always seemed to take up too much of his time. Financial concerns forced some editorial decisions (number of pages per volume, size of pages, and the number of illustrations), and the interrelationship between the *Journal* and the College complicated matters.

As an organ of the College, the *Journal* was supposedly supervised by a publishing committee elected by the membership. During the *Journal's* existence this committee exerted varying amounts of control over policies. When Carson served as editor, the publishing committee had some influence, although it rarely meddled in his day-to-day work.[27] During Procter's years as editor, the publishing committee diminished in significance, perhaps because of Procter's resourcefulness. Near the end of his editorship, Procter admitted at a meeting of the College that he had written all the publishing committee reports since assuming editorial control.[28]

In general the College allowed Procter to do as he pleased with the *Journal*. Only in the financial area did the College seriously hamper him. For example, in 1856 the Board of Trustees decided to provide a free subscription of the *Journal* to each full member of the College. This action hurt the *Journal's* fiscal stability, for it eliminated about one hundred reliable financial supporters just before the panic of 1857. This decision, perhaps a consequence of the enthusiasm of the 1850s boom, constantly irritated Procter throughout the rest of his term.[29]

Economic conditions were difficult for the rest of Procter's editorship, except for a brief respite during 1859 and 1860. As bad as the 1857 panic that had crushed scores of periodicals, the Civil War took Southerners off the subscription rolls of Northern journals.[30] No exception, the *American Journal of Pharmacy* lost about three hundred subscribers.[31] A peaceful man, Procter lamented that since "this Journal is purely scientific and wholly disconnected with political objects, and solely humanitarian in its tendency, no possible objection can be urged against its circulation [in the South]."[32] The loss of the Southern readers especially hurt, inasmuch as few of them belonged to the College and had been paying for their issues. Procter pleaded with his subscribers in the North to pay up their arrears, but with little apparent success.[33] Moreover, in the weeks before the shelling of Fort Sumter business activity slowed to a "dulness" that few had seen before.[34] The combination of declining advertising revenue and fewer paying subscribers forced Procter to go to the College for monetary assistance.[35]

As the Civil War continued, so did the financial difficulties of the *Journal*. Paper prices skyrocketed, along with labor costs.[36] To meet this challenge, the College authorized an increased subsidy to the *Journal*.[37] By cutting back on the size of each number and costly extras like illustrations, Procter brought the accounts of the *Journal* back into black ink before the war's end.[38]

Although the return of Southern subscribers helped the *Journal* in the late 1860s, rampant inflation curtailed Procter's ability to publish articles with illustrations.[39] These financial troubles wore down Procter's enthusiasm as editor and influenced his decision to step down.

Procter's reputation as an editor, of course, did not rest on his achievements as business manager. For twenty years he provided his readers with the best in original and reprinted articles on aspects of pharmaceutical science and art. Moreover, Procter's editorials espoused goals for American pharmacy that formed a critical framework for the progress of the late nineteenth century. He brought to his position all the necessary requisites for success: pharmaceutical skill and knowledge, a huge capacity for work, and an unshakable vision of what pharmacy should become. His predecessors had successfully molded the *Journal* into a conduit for pharmaceutical knowledge. Primarily physicians, these early editors concerned themselves with the construction of a periodical dedicated to the expansion of pharmaceutical science in America. Procter, agreeing with this objective, added his own: to aid the progress of American pharmacy. Maturing in an era when the idea of progress achieved prominence, Procter emphasized this concept throughout his career. His own changing views of pharmaceutical progress were reflected in that extension of his psyche, the *American Journal of Pharmacy.*

Procter's editorials epitomize his journalism, but the selection and editing of copy formed the foundation of his twenty-year labor. Procter had two main objectives for the *Journal:* increasing the amount of original contributions and the practicality of the whole. He accomplished both aims, at least partially. In 1847, before Procter joined Carson as coeditor, the *Journal* contained about thirty-three percent original material (articles, notes, reviews, inaugural essays, and College reports) compared to about sixty-six percent reprinted from other periodicals. Most of the latter came from foreign journals, principally the *Pharmaceutical Journal* and the *Chemical Gazette* of Britain, and the *Journal de*

Pharmacie of France. Even though original articles had made up only about a quarter of the *Journal*'s early volumes, the papers were usually of good quality, often by men of reputation, like Robert Hare.[40]

In Procter's first full year as editor, about thirty-eight percent of the pages of the *Journal* held original writings, but this small increase came mainly from Procter's own editorials and book reviews. As the *Journal* expanded from 392 pages to 576 pages during the 1850s, the ratio of original to reprinted material continued at roughly 35:65.[41] Only in the postwar years did Procter reverse this imbalance, bringing the ratio of original to copied articles to about 50:50.[42]

At first glance there is no thread connecting what Procter put in his journal. It seems to be a potpourri of whatever caught his eye in various scientific periodicals. Viewed broadly, however, the diversity appears related to Procter's pledge to increase the practicality of the *Journal* without sacrificing its scientific standards. He saw no conflict between "pure" science and "applied" science, but instead expressed an eighteenth-century European tradition of the pharmacist as an exponent of the natural sciences. Although he reprinted articles on theoretical subjects, the vast majority of reprinted pieces illustrated science solving practical problems. Whether describing a new chemical method for raising bread dough, an improved varnish, better manures, botanical expeditions to the Amazon, or the research of Louis Pasteur, William Henry Perkin, or Alexander Crum Brown, an article reprinted in the *Journal* demonstrated useful discoveries and developments that arose from chemical and botanical investigation. Procter realized that a small minority of American pharmacists had the time and ability to pursue pharmaceutical investigation. He sought to make his readers aware that their work behind the counter connected with the larger world of scientific progress.[43]

Purely practical or technological innovations were abstracted by Procter and put into "Gleanings." He wanted his readers to aspire to more systematic investigation and to publish their results in the *Journal*.[44] The appearance of John M. Maisch on the pharmaceutical scene in the late 1850s provided Procter with access to the German pharmaceutical literature. For example, in the first number of volume 33 (1861), Maisch began a series of translations of Rochleder's *Proximate Analysis of Plants*.[45]

During his first few years as editor, Procter published almost every sound original contribution that crossed his desk. Even when he disagreed completely with the conclusions reached by an investigator, Procter published the article with his own comments in footnotes.[46] This practice sounds lax, but Procter put a great deal of effort into the replication of scientific investigations reported to his *Journal*. Procter was not an especially inventive investigator by continental standards, but he sought constantly to refine the methods of pharmaceutical science and art in the United States. By checking the work of contributors to the *Journal* when it appeared faulty, Procter helped to keep pharmaceutical science above the vitriolic polemics that plagued American medical science during the nineteenth century.[47]

This open editorial approach allowed some unusual papers to reach print. For instance, in 1853, four pages described a concretion of hairs found in the esophagus of an ox.[48] During the early 1850s Procter did not have total control of the *Journal*'s contents, and the publishing committee or the College as a whole could direct that a certain paper or lecture be published. As a young editor Procter probably had no objection to this practice, because he needed copy. But as the *Journal* grew in reputation, so did Procter's power, and the number of borderline papers declined. The appearance of the *Report* of the Philadelphia College of Pharmacy Alumnae Association in 1865 helped Procter by providing another outlet for popular school lectures and papers.[49] As his editorial control increased, Procter gradually expanded the quantity and quality of original articles in the *Journal*. As he did so, the circulation of the *Journal* rose.[50]

Part of the *Journal*'s survival and eventual growth can be attributed to Procter's own exemplary articles, the investigations of his friends and former students, plus Procter's continuous stream of editorials calling for more original contributions. His own writings in the *Journal* spanned thirty-five volumes and several aspects of pharmaceutical science and art. During the 1850s especially,

Procter would fill empty pages of the *Journal* with his own varied investigations, using different bylines, perhaps to screen his dominance of the *Journal*. The infant American Pharmaceutical Association expanded his contacts with men of similar scientific interests, such as Edward R. Squibb, Frederick Stearns, and Israel Grahame. Philadelphia friends like Alfred B. Taylor, Charles Bullock, Franklin Bache, and Edward Parrish wrote for him. As his students became established, they began producing quality research papers (for example, Albert Ebert and C. Lewis Diehl). Above all, his relationship with John M. Maisch supplied the *Journal* with a steady stream of original investigations and notices.

By the time of the Civil War Procter had carefully molded the *Journal* into a balanced mix of practical and theoretical information dealing with pharmacy and the pharmaceutical sciences. If he disagreed with an author, Procter would indicate his objections in a footnote and invite his readers to investigate the problem themselves and write up their findings.[51] His efforts to stimulate investigations by his readers continued into the editorial department. Whether he discussed pharmacy practice (pharmaceutical organizations, poison dispensing, pharmacy laws, interactions with physicians, or tips on technical manipulation), pharmaceutical education, pharmacopoeial revision, or serious scientific controversy, Procter repeatedly asked his readers for comments, notes, abstracts, or full-length papers.[52]

In the postwar years, when the *Journal* achieved the character Procter had envisioned, some readers criticized a perceived theoretical emphasis.[53] They pointed to practically oriented publications, mainly the *American Druggists' Circular and Chemical Gazette,* which published more formulas and helpful hints for making drug products. Perhaps growing a bit weary of such criticism, Procter reacted strongly: "The apothecaries who work out these [formulas] *are, in general, not . . . liberally disposed;* on the contrary they make specialities of their recipes on all occasions where there is a prospect of pecuniary return." Procter even suspected the motive of some who did not keep their formulas secret: "The [formulas] are contributed by practical men of all grades. One writes to direct attention to a *patented* process; another describes a substance, the

THE

AMERICAN

JOURNAL OF PHARMACY,

PUBLISHED BY AUTHORITY OF THE

PHILADELPHIA COLLEGE OF PHARMACY.

EDITED BY

WILLIAM PROCTER, Jr.

PUBLISHING COMMITTEE FOR 1868;

CHARLES ELLIS,   ALFRED B. TAYLOR,
Prof. JOHN M. MAISCH,   Prof. EDWARD PARRISH,
          AND THE EDITOR.

VOLUME XL.

THIRD SERIES, VOL. XVI.

PHILADELPHIA:
MERRIHEW & SON, PRINTERS,
No. 243 Arch Street,
1868.

Typical title page of the journal that Procter edited for twenty-one years (1850–71).

origin of which is involved in doubt; a third gives a recipe, with the manipulation omitted, as an advertisement; whilst a fourth—a *rara avis*—tells the whole story, and gives the true working formula or process."[54] In typical Procter spirit, however, the editorial called out to all practically minded pharmaceutists with high ideals to submit their own discoveries, notices, and tips for publication: "Let us join hands, then, in contributing more practical articles to our *Journal,* at the same time that it retains its standing as a record of pharmaceutical science."[55]

Spurred by criticism of his editorial policies, Procter renewed his efforts to get more practical material of good quality into his periodical during the late 1860s. His requests did produce responses.[56] Almost all of these authors were practicing pharmacists, with little or no formal scientific training.

In the years just before he handed over the editorial reins to John Maisch, Procter felt satisfied that his efforts had produced a journal "devoted to the advancement of pharmaceutical knowledge, including practical pharmacy, materia medica, chemistry in its general and applied relations, and to the collateral sciences, botany, mineralogy, zoology, &c."[57] Procter left out matters strictly medical. When extracting from medical journals he excluded case reports because they did not deal directly with the pharmaceutical aspects of the new medication discussed.[58]

The American pharmaceutical journal from which Procter usually extracted articles was the *Druggists' Circular,* the only serious national competition for the *American Journal of Pharmacy.* Although this trade newspaper contained more practical and commercial material than the *Journal,* the basic approach of its long-time editor, Dr. L. V. Newton, differed little from that of William Procter, Jr.[59]

A typical postwar issue of the monthly *Druggist's Circular* (sixteen pages of regular issue and an eight-page supplement) included about sixteen pages of advertisements, seven pages of articles, and one page of current market prices. Typifying the *Circular*'s trade emphasis were the advertisements that filled the first few pages of each issue and other advertisements sandwiched between text. The opening reprint section did not stress the crassly commercial, but contained pertinent scientific and technological news. In an 1866 issue, for instance, this section held articles on the history of alum, the vermifuge properties of pumpkin seeds, and remedies for rheumatism.[60] Together with these full-length pieces were abstracts covering a new assay for gold, increased production of Liebig's Extract of Meat, and the best ways to ventilate sewers. Another department in this issue reprinted general scientific articles or abstracts embracing such diverse subjects as aerial locomotion, the analysis of fossils in coal, the philosophy [physics] of a top, and the function of leaves. Strikingly absent are such standard trade themes as how to "buy cheap and sell dear" or "get the most out of your clerks." Newton usually limited thoroughly practical concerns to the "Notes and Queries" department, which served as an open forum for pharmaceutical discussion.

Newton's editorials illustrate his idea of good pharmaceutical journalism. Even more than Procter, Newton stressed the importance of pharmaceutical education.[61] In later years, Newton's editorials shrunk in size and frequency. This contrasts with Procter, who became more vocal and critical as the years went by.

The editorial departments of the *American Journal of Pharmacy* from 1850 to 1871 tell us more about William Procter, Jr., than any other surviving source. Pressed for time by other commitments as professor, pharmaceutist, organizational leader, scientist, and family man, Procter wrote editorials rapidly, apparently with few changes aside from grammatical proofreading. He would sometimes come to the end of the available space, apologize, and pledge to continue discussion of that topic in the next number. His straightforward approach to matters of practice allowed him to speak with direct force to his readers.

Though the editorial form had a long history in periodic literature, Procter arrived in the midst of a "great age of the editor and of the editorial," when editors had almost total control of their publications.[62] Procter probably modeled his editorials after those of the English pharmaceutist Jacob Bell. In complete control of the *Pharmaceutical Journal,* Bell used this platform to rally support for his cause: the elevation of pharmacy's status in England. Procter may have argued to Carson that the analogous position of American pharmacy necessitated a similar platform. At any rate, the editorial department came to life when Procter joined Carson as coeditor.

The early editorial department (1848–49) held writings from both coeditors, but primarily from Procter.[63] In addition to usual editorial subjects, the new department contained obituaries and book reviews.[64] When under the purview of both Carson and Procter, editorials emphasized internal matters (the journal or the College), together with ethical practice considerations.[65] The new drug law regulating the quality of imported drugs attracted special attention.[66]

When Procter assumed sole editorship he shifted the emphasis of the editorial department toward practice issues, such as poisonings from prescriptions, taxes, and drug quality.[67] When the incidence of drug adulteration led to the convention of pharmacists in New York in 1851, then to the founding meeting of the American Pharmaceutical Association in 1852, Procter's editorials played a significant part in the success of both meetings (see chap. 7). During the next twenty years, these editorials reflected the changing scene of American pharmaceutical practice and science as viewed by Procter.

In the early 1850s Procter's editorials dealt principally with efforts to increase drug quality and the status of pharmacy through organization and improved practice. The principal vehicles for these advancements were to be the *United States Pharmacopoeia,* the young American Pharmaceutical Association, and local pharmacy organizations.[68] In addition to combating adulterated drugs and medicines, Procter urged his colleagues to renounce quackery and nostrums.[69]

During the mid-1850s (1854–57) Procter's view widened. He joined the scientific Cavendish Society and gave notice to its publications in the editorial department. Always an Anglophile, Procter commented on the progress of pharmacy in Great Britain and the colonies.[70] At the same time, he doubled his efforts to get more American pharmaceutists to probe the indigenous materia medica for new remedies.[71] Procter assigned more space to book reviews during this period, coinciding with an upsurge in American publishing. In his reviewing he tried hard to provide a balanced viewpoint. For example, historian Alex Berman called Procter's reviews of neo-Thomsonian and eclectic books, "among the very few dispassionate critical analyses of Botanicism to appear in nineteenth century literature. Not being a physician, Procter carefully refrained from passing judgement on the medical and clinical pretentions of Eclecticism, confining his critical analysis to fields in which he was eminently competent to speak, namely, pharmacy, chemistry, and materia medica."[72]

The principal theme of his editorials still remained the benefits organizing could bring to the business and reputation of the individual pharmacist. The issues of the *Journal* just before and after the annual meeting of the American Pharmaceutical Association always carried an editorial either previewing or describing the proceedings. Procter's editorials aided the founding and survival of the young association.

In the years before the Civil War, Procter paid special attention to the American Pharmaceutical Association and particularly to local organizations, which struggled during economic and social distress. For the practitioner, increased competition for decreased business strained the youthful ideals of cooperation; for national leaders, the "impending conflict" distracted attention from organizational priorities. Although Northerners dominated the American Pharmaceutical Association, like similar professional and scientific societies in the United States, several Southern pharmaceutical associations bolstered the national organization.[73] Procter's editorials in the late 1850s stressed the unifying by-products of occupational interaction and interregional botanical research.

After a pause during 1857, new books swamped Procter's desk, with reviews in the prewar years dominating several editorial departments.[74] Procter used book reviews as a medium for the advocacy of self-improvement. By building a personal library and reading, any pharmaceutist could elevate his scientific expertise, enhance his practice, and thereby raise the stature of his profession. As new titles multiplied and gaps were filled in the pharmaceutical literature, Procter became more discriminating in his evaluations.

The Civil War produced a shift in Procter's editorials. War subjects occasionally crept into this department, but Quaker Procter concentrated on other matters. He could not ignore, however, a series of revenue laws generated by the war that directly affected pharmacy practice. One law heavily taxed alcohol, raising its price threefold. Under this tax burden less conscientious pharmaceutists found it hard to follow the many pharmacopoeial formulas that called for the evaporation (and loss) of large amounts of alcohol. As a champion of pharmacopoeial authority and uniformity, Procter battled the high alcohol tax throughout the war and after.[75] With the progress of American pharmacy at a standstill during the war, Procter turned his gaze toward Europe and pharmacy there.[76]

In the postwar years Procter continued to comment on the international pharmacy scene. In 1867 he left America and the *Journal* (for two issues) to make the grand tour of Europe and attend the International Pharmaceutical Congress in Paris. This trip had a profound impact on Procter that was reflected in editorials subsequent to his return.

Before his trip Procter had contended that pharmaceutical progress in America could be achieved through organizing on the local and national levels. Local societies would foster cooperation and education, while discouraging destructive competition; the American Pharmaceutical Association would raise the general status of pharmacy. Outside interventions through laws and regulations were unnecessary. His experience in Europe, however, let Procter step back from American pharmacy and see it in perspective. He compared pharmacy practice in laissez-faire England and regulated France and Italy. The higher standing of continental pharmaceutists impressed him, as did their scientific achievements. Their practice contrasted sharply with that in the United States, where a large expansion of mass-produced medicines and weak social control over the practice of pharmacy allowed almost anyone to open a drugstore.

Procter came back a changed man. He had gone to Europe with the misconception that the distance between the levels of pharmacy practice on the Continent and in the United States had diminished. He found that it was not so and felt disheartened.

Procter's editorials turned more critical. He began pointing out deficiencies in American pharmacy: the lack of a standardized educational system, the decline in scientific investigations by practitioners, and the need for laws regulating the dispensing of dangerous drugs.[77]

Before his European trip, Procter occasionally defended the foibles of the common practitioner. After his trip Procter grew more impatient with the incompetence of many Americans claiming to be pharmaceutists. Another poisoning by prescription in 1869 caused him to comment that "in view of the yet imperfect education of many to whose lot dispensing falls, . . . physicians should make their prescriptions so plain as to be read by even the *mediocre.*"[78]

Instead of holding Procter back, this new realism released him from old illusions and misconceptions that had constrained him. He went into the 1870s more open to change. For example, he favored adoption of the metric system in the *Pharmacopoeia* and advocated the passage of laws to regulate pharmacy directly, two nearly radical positions for the time.[79] He poured his energies into organizational work on the local level in an attempt to stop the intellectual decline he saw threatening American pharmacy as a by-product of the rise of factory-made medicines and destructive competition. He again called for more practice-oriented material in the *Journal*. He hoped that if pharmacists shared practical experiences they might continue to manufacture drugs themselves.[80]

Naturally a conservative man, Procter stood his ground on several old and important issues. During his last year and a half as editor, he continued to champion the progress of pharmacy through cooperation and education. Strong-minded on the subject, he rejected the new pharmacy degree awarded by the University of Michigan because no practical experience was required.[81]

Still enthusiastic about American pharmacy's future, Procter wearied of the struggle to keep the *Journal* solvent. He grumbled that "there are quite a number of our subscribers whose practice conveys the impression that printers and paper makers work gratuitously."[82] He also tired of begging for financial support from the College. In 1869 Procter vented some of the anger and frustra-

Tenth Street buildings of the Philadelphia College of Pharmacy, where Procter edited the *American Journal of Pharmacy* from 1868 to 1871 and taught practical pharmacy from 1872 to 1874. The College lecture halls were in the building on the left, the office of the *Journal* in the smaller building to the right. (From Joseph W. England, ed., *First Century of the Philadelphia College of Pharmacy, 1821–1921* [Philadelphia: Philadelphia College of Pharmacy, 1922], 147.)

tion that had built up. He revealed that the College publishing committee had been a dead letter since the early years of his editorship and that he himself had carried on its work and composed its yearly reports. Procter hoped that the College would understand the difficulties and help him to increase subscriber payments.[83]

Other nagging troubles at the *Journal* can be discerned in this unusually long publishing committee report. Original contributions to the *Journal* from College members had decreased, which Procter blamed on "greater absorption by business cares."[84] This situation forced the editor to spend more time searching European journals for investigative reports. Many were in French, which Procter had to translate—again increasing his work load. The German journals offered another source, but Procter could not translate German. Soon after this report, he "entered into a business arrangement with Prof. John M. Maisch, by which each number of the *Journal* [would] contain a regular contribution of translations from that source." Thus began the formal involvement of Maisch, Procter's editorial successor, in the affairs of the *American Journal of Pharmacy.*[85]

Maisch had been writing original articles and abstracts of German articles for the *Journal* for many years before 1869, and Procter had been impressed by the "numerous valuable original contributions" he received from Maisch.[86] Maisch had succeeded Procter as professor of practical pharmacy (although only for one year) and demonstrated excellent scientific skills. Urdang has contended that Procter purposefully held control of the *Journal* until he was sure that the editorship would go to Maisch, not to Edward Parrish.[87] Although the evidence for this is circumstantial, Procter's obituary of Parrish does reveal an ambivalent opinion of Parrish's skills. He saw Parrish as the consummate practical apothecary, but as far below Maisch in scientific knowledge and ability.[88]

Before turning over control to a new editor, Procter took several bold steps to get the *Journal* on a better financial basis. With the assent of the publishing committee, Procter initiated in mid-1870 a monthly advertising sheet to be sent to subscribers, which would include pharmaceutical news.[89] Soon afterward, in the editorial that announced his resignation, Procter told of plans to publish the *Journal* monthly and to hire a business manager.[90] These decisions had been approved at a special meeting of the College held on December 5, 1870.[91] When Procter reported the appointment of the new business manager, Henry H. Wolle, he characteristically asked subscribers to pay up their arrears.[92]

In the last installment of his editorial department, Procter followed the pattern that had served him well for over twenty years: *Journal* business (his resignation), a more personal piece (his responses to well-wishers), College news (concerning the new practical laboratory), pharmaceutical organizations or laws or both (in this case New Jersey, Rhode Island, Illinois, and California), a dispensing tip (recipe for hive syrup), reviews *(Yearbook of Pharmacy)*, and obituaries (Eugene Massot, F. A. G. Miquiel, James S. Muspratt, and Ferris Bringhurst).[93]

With obvious relief Procter passed the helm to John M. Maisch, a man whom he respected and trusted. Procter had spoken to a whole generation of American pharmacists, perhaps the first full generation of "true" pharmacists in the United States. ("True pharmaceutist" was a term Procter used to describe those who made the accurate compounding and dispensing of prescriptions from official ingredients their special vocation.) He had spoken like a loving father to his adolescent sons—with a firm, yet understanding tone. He had scolded rarely. Instead, Procter relied on encouragement, on example, and on leadership.

Before the assembled members of the College gathered at the year-end meeting of 1870, Procter reflected on his labors:

It is now thirty-four years since my connection with the American Journal of Pharmacy as a contributor commenced, and about twenty-five years as co-editor and editor. During this period time and labor have been freely given to make the work a continuous record of the progress of Pharmacy at home and abroad. For many years it was a labor of love, and despite the great sacrifice of time occasioned by contributing to its pages, the labor was cheerfully given. Of latter years a change has occurred in this respect; the work has been continued regularly as a matter of duty, but it has ceased to be a pleasure. Under these circumstances, I desire to carry out an intention entertained for several years, and withdraw from the editorship. In order to give the College time to select a successor, I have deemed it best to offer this my resignation at this meeting, to take effect at the annual meeting in March, when the stated time for electing an editor arrives.[94]

With those words Procter left the editorship of the *American Journal of Pharmacy.* He had taken the reins of a mainly scientific journal from Joseph Carson and transformed it into a medium for vocational development. Through its pages he wrote on every aspect of pharmacy and made his name nationally known. When other colleges of pharmacy began publishing their own journals in the 1850s and 1860s, as often as not they copied Procter's style and benefited from his advice. Above all, he demonstrated effectively the use of the editorial in American pharmaceutical journalism. A quiet man and a staid lecturer, Procter let his pen speak with the fire and emotion his voice could not easily express. Along the way Procter helped define the scope of both pharmaceutical science and practice in the United States of the late nineteenth century.

In the next number of the *Journal* after Procter's retirement, an article entitled "Ferrate Elixir of Cinchona" appeared with the byline "By the Editor." Even though the subject was a Procter favorite—iron preparations—the piece came from Maisch.[95] A formative period in American pharmaceutical journalism had come to a close. Twenty years of what Maisch called "fearless management" were over.[96]

6 Pioneering Pharmaceutical Educator

In 1846 the Philadelphia College of Pharmacy selected William Procter, Jr., to fill a newly created chair in practical pharmacy. This action marked a turning point in the history of pharmaceutical education in the United States. Previously, pharmacy schools had expected practitioners to provide practical training in the apprenticeship years or had looked to physicians to teach the subject in combination with the course in the natural history of drugs, materia medica. The establishment of a chair combined with the selection of pharmacist Procter by the Philadelphia College of Pharmacy showed that the young vocation had achieved enough self-confidence and self-identity to insist that practical pharmacy could stand alone as a discipline.[1]

Before one of his graduating classes Professor Procter welcomed the young men into the brotherhood of pharmaceutists with a challenge that epitomized his views on the relative importance of formal education in the development of the American pharmacy:

My friends! We are about to part. . . . [L]et me tell you a truth . . . which twenty-seven years of professional experience have but tended to confirm. The only permanent cure for the evils [that confront ethical pharmacy] is the union of knowledge, skill and integrity in the dispenser of medicines. As teachers, we have endeavored to give the first. The second you have gained in the several establishments of your special preceptors.

124

The last rests with yourselves. It is the crowning qualification, which, as worn or discarded, shall render you a blessing or a curse to the communities in which your future lots may be cast.[2]

Throughout his teaching career, which spanned nearly thirty years, Procter clung tenaciously to the traditional model of pharmaceutical education that placed primary emphasis on apprenticeship and self-study, and that viewed formal education as merely a "finishing." This viewpoint became almost reactionary near the close of his life. Yet it may have been the only choice for Procter, a man who believed strongly in self-reliance and open competition in pharmacy. This chapter examines Procter's significant role in the development of pharmaceutical education in America, showing how his example and leadership both furthered and encumbered the progress that formal education seemed to offer.

Procter probably formed his ideas concerning pharmaceutical education through his own experiences. Throughout the nineteenth century, almost all American pharmacists received their training by a combination of apprenticeship and self-instruction. Apprenticeship nominally took four years, but because no law enforced this requirement, many ambitious pharmacists-to-be ignored it. These young men sought "a minimum knowledge that would permit financial success . . . in pharmacy as a trade."[3] The ideal situation, unattainable by most apprentices, was to attend courses at a school of pharmacy after the bulk of their in-shop training and be "rounded off." Formal education, therefore, was a luxury usually reserved for those who apprenticed at the better shops in the few cities with courses of instruction.[4] Most apprentices received training from pharmacists who knew little and cared less about scientific or ethical pharmacy.[5]

Procter characterized the situation in an address to the American Pharmaceutical Association:

Our country has been deluged with incompetent drug clerks, whose claim to the important position they hold or apply for is based on a year or two's service in the shop, perhaps under circumstances illy calculated to increase their knowledge. These clerks in turn become principals, and have the direction of others—alas! for the progeny that some of them bring forth, as ignorance multiplied by ignorance will produce neither knowledge nor skill. . . . It has been found that there are three classes of individuals engaged in pharmaceutical pursuits . . . to whom particularly this address is directed: First, those who are imperfectly acquainted with pharmacy and are in business for themselves; secondly, those who have been but half educated as apprentices and who are now assistants receiving salaries, having the responsibility of business entrusted to them; and thirdly, those who are now apprentices or beginners under circumstances and with ideas unfavorable to the acquirement of the thorough knowledge of the drug and apothecary business.[6]

As a solution to this problem Procter did not insist on more pharmacy schools, but encouraged preceptors and apprentices to read pharmaceutical literature regularly and to experiment with new methods before using them to compound medicines.

At first, Procter's adherence to the apprenticeship mode of educating pharmacists appears incongruous with his professed faith in education as the machine of progress for American pharmacy. Although it is impossible to tell exactly why Procter believed so strongly in apprenticeship, it is possible to make a few speculations.

Procter received his early education in Friends schools, which stressed the practical and useful over the artistic and frivolous. Theoretical speculation for its own sake was antithetical to the simplicity of life ordained by the Quaker "Disciplines."[7] Moreover, Procter served an almost perfect apprenticeship: his preceptor, Henry Zollickoffer, encouraged him to attend the Philadelphia College of Pharmacy and lectures at the University of Pennsylvania, gave him full use of a laboratory for scientific investigations, introduced him to Elias Durand and his circle of scientific friends, and treated him like an adopted son. Procter, a quiet, introspective youth with great drive, thrived in this environment, laying the foundation for his later achievements.[8] Lastly, the apprenticeship system kept the control of pharmaceutical education in the hands of pharmacists. Many of the pharmacy schools of the nineteenth century, like that of the Philadelphia College of Pharmacy, sprang from local pharmaceutical associations that arose because outsiders, usually physicians, tried to gain control over the training or certification of pharmacists.[9]

After four years of apprenticeship, Procter attended night courses in the lecture hall at the College taught by Franklin Bache (chemistry) and George B. Wood and Joseph Carson (materia medica).[10] (The course in chemistry was a general course that also included information on the sources and properties of inorganic drugs. Materia medica dealt with the natural history of galenicals, their commercial sources, organoleptic tests of quality, their physical properties, and their medical uses.)[11] Like medical students, pharmacy students sat through the same lectures twice. "Since it was assumed that the novice of pharmacy learned his art best in the apprenticeship, and because of the limited instruction time available, pharmacy as academic instruction [was] part of the catch-all course called materia medica."[12]

Physicians had taught these two subjects since the beginning of instruction at the College. They were the logical choice because of their experience and education, which included a small amount of training in pharmacy.[13] The physician professors inserted some pharmacy in their lectures on chemistry and materia medica.[14] Pharmacy proper, however, was not seen as an academic subject and did not enter the College curriculum until 1847.[15]

At the semiannual meeting of the Philadelphia College of Pharmacy on September 5, 1845, a statement was read, signed by Procter, A. J. Duhamel, and Edward Parrish, accompanied by a resolution that "a committee of nine members be appointed to take into consideration the propriety of creating a new professorship, the occupant of which shall be called 'The Professor of Theoretical and Practical Pharmacy'—and if they deem it expedient, to mature a plan for the consideration of a future meeting."[16]

To the assembled members of the College, Procter and his young friends insisted that the great variability of apprenticeship experiences had caused many candidates for graduation to be poorly prepared for examinations in practical and academic subjects. Even though they learned materia medica and chemistry adequately from lectures, the students needed pharmacy instruction.[17] Reflecting the uncharted ground of the proposed position, Procter observed that "the Professor of Pharmacy . . . must enter a field of labor scarcely less extensive than that of either of his

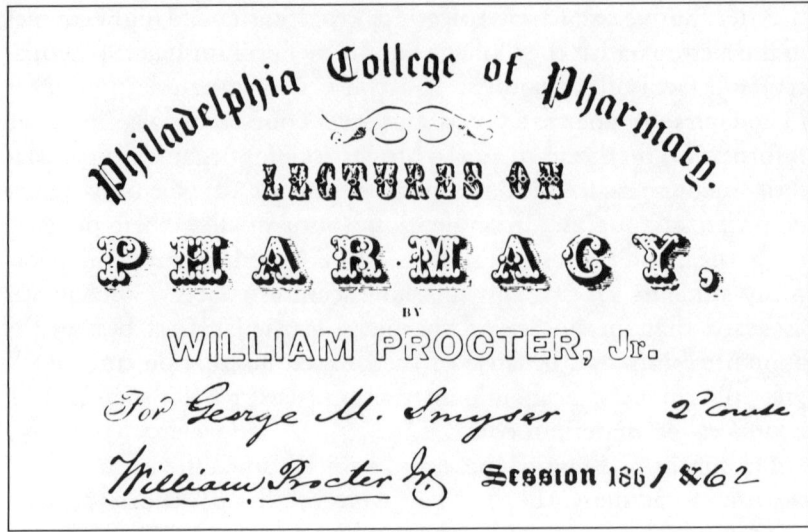

It was through the sale of such tickets (about ten dollars each) that professors received their monetary compensation for teaching. (From *American Journal of Pharmacy* 113 [1941]: 335.)

colleagues in the school, and one which he will have to traverse in the double capacity of teacher and learner."[18] The arguments of Procter, Duhamel, and Parrish proved convincing, and after "an animated discussion, the resolution was adopted."[19] At a special meeting of the Board of Trustees held on June 1, 1846, the committee made its report and on its recommendation the board unanimously selected Procter as the College's first professor of practical pharmacy.[20]

In his introductory lecture, Procter enunciated views on the pharmaceutical profession that he would hold, relatively unchanged, for twenty-five years. Although he mentioned the importance of attending lectures, Procter emphasized that students could learn practical pharmacy only through the apprenticeship process.[21] Even to Procter the young subject of practical pharmacy took a place secondary to the well-established disciplines of materia medica and chemistry.[22] Procter praised the high level of expertise German pharmacists achieved under educational requirements imposed by the state, but contended that open competition in the United States offered more opportunity to young men.[23]

Procter saw his mission as twofold: first to demonstrate the rudiments of practical pharmacy to his students with the hope that they would replicate his examples; then to describe the finer points of pharmacy practice, revealing the pitfalls that threatened the inexperienced practitioner. These challenges included chemical incompatibilities, drugs of poor quality, and the lack of recognized standards for drugs and medicines. Procter sought to connect the theoretical and practical aspects of pharmacy, tying together what the students learned in chemistry lectures with what they experienced in the shop.[24] Through his two-pronged efforts—in the classroom and in his shop laboratory—Procter laid the foundations for the distinct discipline of pharmaceutical chemistry in the United States.

Procter chose Franklin Bache as his primary professorial role model. In his diary he wrote, "Dr. Bache deserves the greatest credit for the considerate manner in which he discharges his duty to his students."[25] He respected the organization and clarity of Bache's lectures.[26] Procter and Bache had similar reputations as teachers: kind, considerate, knowledgeable—and boring.[27] A pharmacist once commented on the anticipation he felt waiting for the renowned Procter to lecture. He had come all the way from St. Paul, Minnesota, to attend the course at the Philadelphia College of Pharmacy. Procter stepped up to the front of the lecture hall and began. At first, the new student sat disappointed, but later became impressed by Procter's careful demonstrations and the depth of his learning, remarking, "I immediately recognized in him the foremost representative of the art and science of pharmacy."[28]

Joseph Remington remembered his former teacher:

As a lecturer William Procter did not shine with the same brilliancy that he did as a writer. He never sought to brighten any subject by flashes of humor or a digression from the subject in hand. . . . It must not be supposed that Procter's lectures were uninteresting; they were full of meat, and those who wanted to learn had no difficulty whatever in absorbing knowledge. He had a systematic and orderly way of grouping facts, and his lectures were well illustrated with experiments; he always took hours to prepare his experiments beforehand. . . . But it was after the lecture when the charm of his personality was realized. There was always a knot of students collected at the end of his counter to ask questions. The painstaking care by which he sought to remove all difficulties, the pleasant smile upon his countenance, and his parting question to the student, "Are you sure that you thoroughly understand?" gave him a power and an influence which nothing else could. Procter could not be called a fluent lecturer, but he was a born teacher, and never failed to command the highest respect from the students.[29]

For the organization of his course, Procter turned to the writings of the French pharmacist Eugène Soubeiran.[30] Procter resisted the temptation to go through the *Pharmacopoeia* from beginning to end, a simple but confusing way to discuss pharmacy; instead he grouped together drugs "based on the similarity of active principles." This arrangement allowed him to relate pharmaceutical compounding to the emerging science of medicinal chemistry.[31]

Skeptical about the value of a lecture course in the teaching of a practical skill, Procter once wrote:

Mere lectures will not make apothecaries, but oral instruction properly illustrated by [demonstration] experiments and diagrams, when addressed to earnest young men or boys engaged in the daily routine of the shop, is productive of the highest usefulness. It corrects their crude notions of the phenomena they meet with; it suggests improvements in shop practices followed from mere tradition, and it opens out before them as an illustrated map, the length and breadth and the capabilities of their profession, as a sphere of usefulness and a field of ambition.[32]

During Procter's years at the College of Pharmacy, lectures were held three times a week in the evenings from the beginning of October to the end of February. Students interested in graduating were required to attend the course of three subjects twice. Unlike medical school, which was usually a daytime affair, pharmacy classes took place at night, after the apprentices had worked all day in the shop. Thus a group of tired young men, many forced to attend lectures by their employers, confronted Procter and his colleagues. Procter may have relied heavily on demonstrations because his speaking style failed to keep the attention of his class. Popular teachers, like the flamboyant Edward Parrish, rarely used such props.[33]

Demonstrations dominated the first quarter of Procter's course, which discussed the basics of shop manipulations, such as weights and measures, sources of heat, trituration, clarification, decoloration, solution, maceration, infusion, decoction, percolation, vaporization, distillation, precipitation, and neutralization. For example, on the subject of making emulsions Procter got down to specific detail by advising that "the pestle should be held between the thumb and two first fingers of the right hand, in such a manner that whilst the operator holds it with sufficient firmness to give the requisite rotatory motion, it has also a swinging motion as if in a movable socket."[34]

After reviewing the basics of pharmaceutical practice, Procter spent the remaining three-quarters of his course discussing drugs and preparations. He began by covering the products associated with common proximate principles of plants such as lignin (cellulose), starch, gums, volatile oils, and resins. Procter explored each area broadly. For example, under lignin, he lectured on several different acetates because most acetic acid came from the breakdown of wood. Alcohol, ether, and chloroform fell under the heading of derivatives of sugars.[35] For most drugs Procter briefly outlined their source, basic chemistry, constituents, and tests for purity before getting to the key point of the course: the best methods of preparation. He stressed assay methods for certain potent plant drugs like quinine and morphine, while emphasizing purification techniques with mineral drugs like calomel and tartar emetic.[36]

In comparison with the teachers of materia medica and chemistry, Procter was at a disadvantage. During the mid-nineteenth century the maturing discipline of chemistry continued to make interesting discoveries that held the attention of students and showed the value of science. The materia medica professor could combine lectures on new drugs with tales of ancient herbs and exotic lands. Procter's subject, practical and theoretical pharmacy, held little of the mystery of the past or the potential glory of the future.[37] Instead, when Procter came to preparations he attempted to convey to his students the professional satisfaction received when making the best possible drug product. To Procter, there was no higher goal in pharmacy than the pursuit of the "perfect preparation," one that contained all the virtues and characteristics of a crude drug in a stable, uniform, consistent, and elegant form. He did not realize it, but with this quest Procter had begun the academic discipline of pharmaceutical chemistry in the United States.

In 1866, after teaching practical pharmacy at the College for nearly twenty years, Procter decided to resign his chair, citing increasing demands on his time from his shop, the *American Journal of Pharmacy,* and the American Pharmaceutical Association. Although he did not mention it, his failing health may have contributed to the decision.[38]

The availability of a competent replacement, John M. Maisch, further influenced Procter. He had long admired Maisch for his "untiring assiduity and . . . ability."[39] After a year of teaching practical pharmacy Maisch exchanged chairs with Edward Parrish, who had been teaching the materia medica course since 1864.

Six years after Procter resigned from teaching, distant fighting on the Plains between Indians and white settlers indirectly led to his return to the classroom. As a leader of the Friends community in Philadelphia, Edward Parrish volunteered as an intermediary between the warring factions in the West. Parrish traveled out to Fort Sill in Kansas as part of a government delegation. On the way he became violently ill from a "malarial fever" and died soon after arrival at the fort on September 9, 1872.[40] The news shocked the College, which recognized Parrish as one of its stars. Caught without a professor for the practical pharmacy course that was scheduled to begin within a month, the trustees turned to Procter. He accepted their call with reluctance.[41]

Procter got out his old lecture notes and returned to the classroom. The first year back went well. He enjoyed teaching, but the walks home late at night in the winter months undermined his weak constitution. Because of his deteriorating health, Procter planned to resign at the end of the 1873–74 course.[42] On February

10, 1874, just before the end of the term, Procter died a few hours after giving a lecture.[43]

Few American pharmacists took formal courses like those offered by the College throughout the nineteenth century, and far fewer actually graduated.[44] Procter figured that only one-twentieth of American pharmacists had attended any lectures dealing with pharmacy.[45] Even though only a small fraction of American pharmacists heard Procter lecture, he had significant influence through his educational publications, *Practical Pharmacy* and the "Syllabus of a Course of Study . . . of Pharmacy."

When Procter began teaching at the College one glaring inadequacy revealed itself: the absence of an appropriate textbook.[46] Although several good chemistry texts existed, Procter had no English-language manual for pharmacy at his disposal.[47] He did not have to wait long, however, for in 1847 Karl Friedrich Mohr published his *Lehrbuch der pharmaceutischen Technik,* which, when translated and adapted, served as the basis for pharmaceutical textbooks in the English-speaking world for years to come.

Characterized by Urdang as a "milestone" in the development of the pharmaceutical textbook, Mohr's *Lehrbuch* offered British and American pharmacy instructors just what they needed: a comprehensive guide to laboratory and dispensing techniques and apparatus. Mohr wrote his manual as a handbook for the increasingly complex laboratory manipulations that an *Apotheker* was called upon to undertake.[48]

Mohr was perhaps the most qualified man alive at that time to compose such a volume. Son of a well-established German pharmacist and chemical manufacturer, Mohr was groomed to run the family operation. He received the best possible scientific education and rose above his frail physical condition to the highest levels of chemical and pharmaceutical expertise. Perhaps best known for his masterwork, *Lehrbuch der chemisch-analytischen Titriermethoden* (1855–56), Mohr concentrated on pharmaceutical matters in his earlier work.[49] His familiarity with the difficulties attending many pharmaceutical preparations led him to explore possible solutions. Along the way he invented or improved much of the standard chemical and pharmaceutical apparatus of the mid-nineteenth century. Whether he was simplifying methods for specific gravity determination or improving the cork borer, Mohr's inventive genius shines through.[50] The *Lehrbuch* served as a showcase for his technological innovation.

The restricted nature of Mohr's book appealed to the English pharmaceutical chemist Theophilus Redwood. A teacher of pharmacy like Procter, he too had searched for an appropriate classroom text.[51] When Redwood first received a copy of the *Lehrbuch*, he intended simply to translate it into English with a few additions. To "meet the requirements of English Pharmaceutical Chemists," Redwood eventually added new material that doubled the size of the book. Even though Redwood had received permission to translate the *Lehrbuch* from the German publishers, he felt no compunction to translate Mohr's text directly. Instead he used its excellent illustrations as focal points of each subsection and paraphrased Mohr's comments.[52] In those places where Mohr waxed on about the superiority of his method over another, Redwood simply described the apparatus. For example, Mohr spent six pages on the advantages of his self-supplying cutter for medicinal plants; Redwood copied Mohr's woodcut and described it in a single page.[53]

Redwood followed closely the arrangement of the *Lehrbuch*, rarely omitting sections unless they applied only to Germanic pharmaceutical practice, for example, weights and measures. When he disagreed with Mohr's recommendations, as in the case of filter paper (Mohr preferred round, Redwood square) Redwood provided the reader with Mohr's section, followed by his own opinions. Bowing to the German's reputation and expertise, Redwood did this in an evenhanded, almost apologetic fashion.

Aside from illustrations, Redwood added elaborations on the methods particular to British pharmacy and scientific explanations that Mohr probably felt were unnecessary for his better-educated audience. For instance, Redwood went to great lengths to explain the physics of the syphon; Mohr did not.[54]

Lastly, Redwood rounded out the text as a manual for a practical pharmacy course. In addition to discussing "techniques," Redwood detailed methods of dosage form preparation. This last section, entitled "Aids to Dispensing," was Redwood's major contribution; Mohr left the art of the apothecary to other authors. The finished textbook was about one-half Mohr and one-half Redwood.[55]

In 1848 Redwood's revision of Mohr's *Lehrbuch*, entitled *Practical Pharmacy*, arrived in America. Procter, an admirer of continental pharmacy and of British culture, agreed to edit the American version of the book for the prominent publisher Lea and Blanchard of Philadelphia.[56] Procter saw similarities in American and English pharmaceutical practice and reasoned that a modified version of Redwood's manual would serve well as a "handbook for the American apothecary."[57]

Procter's approach as the American editor of Mohr and Redwood's *Practical Pharmacy* differed significantly from that of Redwood. Following a long tradition of other American editors of foreign scientific works, Procter tried to adapt the European text to American conditions. American publishers asked their commissioned editors to change spellings and add a little local material, but not to rewrite the texts. Usually these editions were unauthorized, given that the United States had no copyright agreements with other nations in the mid-nineteenth century, and American publishers usually paid the original European authors nothing for their work.[58]

In contrast to most American editors, Procter significantly rearranged the text and added 100 illustrations and 150 pages of his own, or about one-third of the book.[59] Redwood had eliminated Mohr's arrangement of chapters, breaking the text up into three major parts, with the main body describing the various pieces of pharmaceutical apparatus in a rather haphazard fashion. For example, Redwood had covered the following topics in this order: "Steam," "Extracts," "Tinctures," "Presses," and finally "Furnaces"; Procter put "Sources of Heat" before "Steam" and discussed other processes in a logical progression. Instead of running all of them together, Procter gave each process or major apparatus a separate chapter.[60]

Even though he wrote almost one-third of the American edition, Procter continued to refer to himself as the editor throughout

PRACTICAL PHARMACY:

THE ARRANGEMENTS,

APPARATUS, AND MANIPULATIONS,

OF THE

PHARMACEUTICAL SHOP AND LABORATORY.

BY

FRANCIS MOHR, Ph. D.,

ASSESSOR PHARMACIÆ OF THE ROYAL PRUSSIAN COLLEGE OF MEDICINE, COBLENTZ;

AND

THEOPHILUS REDWOOD,

PROFESSOR OF CHEMISTRY AND PHARMACY TO THE PHARMACEUTICAL SOCIETY
OF GREAT BRITAIN.

EDITED, WITH EXTENSIVE ADDITIONS,

BY

WILLIAM PROCTER, Jr.,

PROFESSOR OF PHARMACY IN THE PHILADELPHIA COLLEGE OF PHARMACY.

ILLUSTRATED BY FIVE HUNDRED ENGRAVINGS ON WOOD.

PHILADELPHIA:

LEA AND BLANCHARD.

1849.

Procter adapted for American use Theophilus Redwood's translation of
Karl Friedrich Mohr's *Lehrbuch der pharmaceutischen Technik*. *Practical
Pharmacy* was the first pharmaceutical textbook published in North
America.

his career. This reflects a combination of his humility and his approach to the textbook. His own areas of investigation in the 1840s—iron preparations and the analysis of vegetable drugs—did not receive emphasis in his revised text. Procter probably judged the latter as outside the realm of practical pharmacy and the former as too specialized to merit special treatment. The generally poor level of practice by American pharmacists, a recurrent Procter theme, dictated that the basics be stressed.

Many of the additions Procter made to Redwood's text dealt with the special conditions confronting pharmacists in the United States. For example, American pharmacists rarely had access to gas for light and heat, so Procter provided detailed explanations and illustrations of alternative heat sources such as alcohol lamps. In contrast to Britain, where druggists could supply dispensing chemists with a wide selection of chemicals, in America pharmacists had to prepare more of their own basic ingredients. Thus Procter included additional information on rudimentary chemistry, such as techniques of calcination, oxidation and reduction, crystallization, and precipitation. The illustrations of fairly large equipment that accompany this section of the book were probably furnished because many of Procter's classroom students apprenticed with wholesale druggists in Philadelphia and planned to continue with those firms after graduation.[61]

For the benefit of both students and practitioners, Procter added conversion tables for weights and measures. Students, of course, needed to memorize these; practitioners often intermingled the troy and avoirdupois systems when making large preparations. Small drugstore weights were usually troy, up to one pound, which equaled 5,760 grains; for amounts over 2 troy pounds, avoirdupois pounds equaling 7,000 grains were used.[62]

The largest original section by Procter dealt with "General observations on the preparation and purification of the fixed oils and fats," and filled chapter 14. As a free-standing chapter it reflects Procter's style and approach better than any other part of the American *Practical Pharmacy*. In direct contrast to the chatty style of Mohr, Procter's prose reads like a dry lecture committed to print. Starting with a sentence that justifies the chapter, Procter

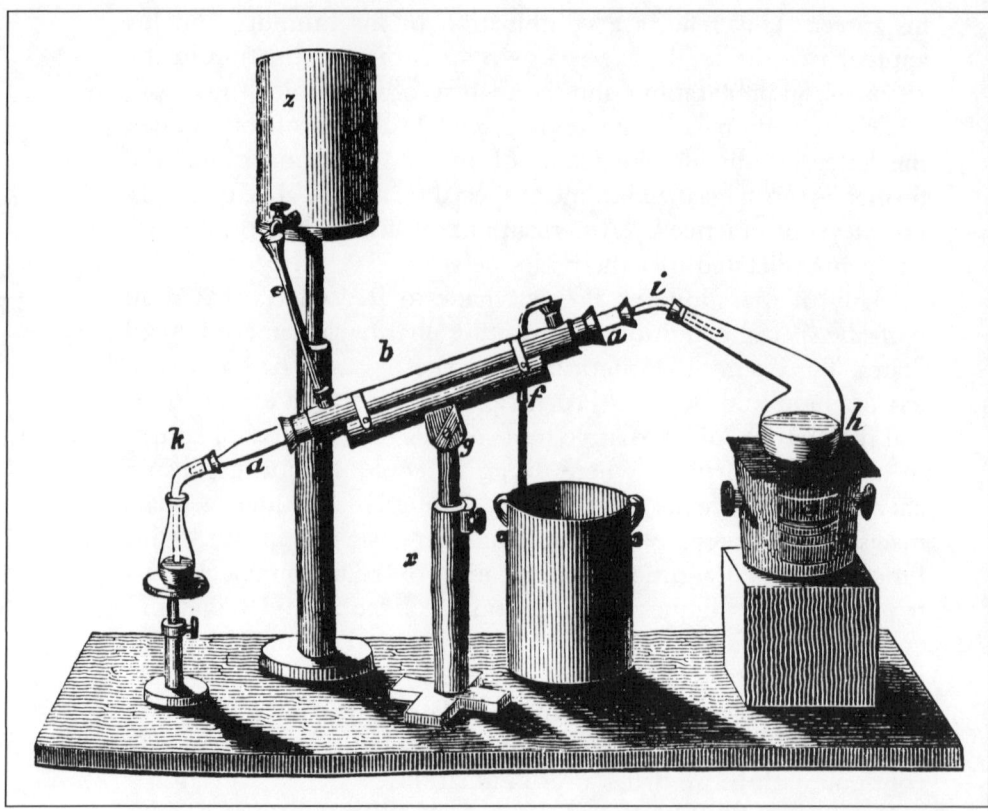

Ideally, when a pharmacist undertook distillation, he had the time and equipment to set up his apparatus in this classic form developed by Justus von Liebig. A glass tube (a) is surrounded by a copper tube (b) resting on a wooden stand (x). The retort (h) is placed over a charcoal fire connected with the condenser by a small glass tube (i). Cool water flows from vessel z, escaping via f into another container. If done properly, the distilled product is conveyed via k into a receiver. (From Francis Mohr, Theophilus Redwood, and William Procter, Jr., *Practical Pharmacy* [Philadelphia: Lea and Blanchard, 1849], 323.)

goes from the general to the specific, eventually detailing individual processes. The emphasis is placed on the advantages of one process over another, especially concerning the potency and appearance of the final product. This chapter, however, differs from others in that Procter casually refers to local conditions and practitioners in Philadelphia.[63] Two of the most important pharmaceutical preparations of the mid-nineteenth century take up several pages: mercurial ointment and lead plaster.[64]

Procter's efforts to make this book American are perhaps best illustrated by his comments on extemporaneous pharmacy. On the mercurial preparation blue mass he contends that owing "to the high price of mercury, and the competition of manufacturers, many instances of adulteration have occurred, either by substituting some heavy substance, as blue slate for the larger part of the mercury, or by simply increasing the amount of conserve at the expense of that of the metal."[65] He then gives a rather simple method for determining the mercury content of blue mass. It is not the most accurate, but it does provide an adequate measurement of the active ingredient. Lastly, Procter describes two new machines used in the United States for the production of blue mass.[66]

The final chapter in Procter's edition of *Practical Pharmacy* was totally new and covered apparatus and reagents for testing drug and chemical purity. According to Procter, the pharmacist "although not a professed analytical chemist, and not possessed of the requisite apparatus for conduction of many of the processes required in analysis, is nevertheless frequently called upon, both for his own satisfaction and at the request of physicians, to perform qualitative examinations of substances. The disposition which exists to adulterate commercial chemicals is so great, that he has constantly to be on the alert to avoid being imposed upon."[67] Procter pointed out that the forty-nine reagents necessary for this work were themselves often sold in an impure state. For this reason Procter gave assays for the reagents, not for the pharmaceutical chemicals. To test the latter, Procter suggested that pharmacists refer to chemistry texts or try some of the reagents on pure samples.[68]

In addition to the major changes given above, Procter made scores of small changes in Redwood's book. These usually involved a paragraph or so on American (urban East Coast) conditions or tips that Procter had come across in practice. In these little hints, perhaps, lay the essence of Procter's contribution. He possessed some of the scientific skill of Redwood and a considerable degree of technological acumen, although short of Mohr's. What Procter added to the final version of this international pharmacy textbook was the touch of a masterful practicing pharmacist. Procter had struggled long to make his little business a success, without the family or business connections that Mohr and Redwood possessed. He understood the low state of American pharmacy and saw his book as a tool for self-help. Although he hoped that American pharmacy would someday produce scientists like Scheele and Sertuerner, Procter knew that practical competency had to precede theoretical excellence. In his own work Procter did not shun the theoretical, as American scientists have been accused of doing; instead he saw theoretical science as outside the mainstream of pharmacy, especially as it was practiced in America.[69]

All in all, Procter's book was a milestone just like Mohr's. It was the first purely pharmaceutical textbook published in North America. By making Redwood's version of Mohr's *Lehrbuch* applicable to American practice, Procter pushed American pharmacy a bit closer to independence and self-sufficiency. With a small library composed of the *United States Dispensatory,* Fowne's or Hare's *Chemistry,* and Procter's *Practical Pharmacy,* a young man with a drugstore could practice a fair grade of pharmacy; with the addition of Pereira's *Materia Medica,* a current copy of the *United States Pharmacopoeia,* and a subscription to the *American Journal of Pharmacy,* a man could aspire to a high level of practice with diligent self-study.

By itself, however, Procter's book probably did little to raise the level of pharmacy practice in America. It sold poorly and Lea and Blanchard never reprinted it. Over the years the listing of the book slowly shrank in their advertisements and disappeared altogether when the stock was finally exhausted in the 1870s. Procter's friend Edward Parrish, six years later in 1855, wrote *An Introduction to*

Practical Pharmacy, which sold well and eventually went through several editions for the same publisher. Parrish oriented his book more toward the numerous country practitioners of both medicine and pharmacy, rather than the small urban elite of educated pharmacists interested in Procter's approach. Thus the *Introduction,* the second pharmaceutical text published in the United States, was the first book to reflect the nature of American pharmaceutical practice accurately.[70]

In contrast to the limited influence Procter's *Practical Pharmacy* apparently had on the pharmaceutical community, his model "Syllabus" of 1858 had considerable impact. It arose from a committee that the American Pharmaceutical Association set up to design a syllabus "to be used in connection with the Dispensatory, as a guide book for study, which will take the place of the lecturer . . . that would prove of great service to apprentices and assistants, and be eminently creditable to the Association, under whose auspices it might be published."[71] As chairman of the committee Procter attempted, unsuccessfully, to involve the rest of the committee. He eventually composed the syllabus himself.[72]

In the introduction to his syllabus, Procter expressed opinions about its utility that reveal much about the state of American pharmaceutical education before the Civil War:

When a lad . . . has made up his mind to enter upon the business of a pharmaceutist . . . he should be fully aware . . . of the two-fold nature of the functions it requires to be performed, viz: those of a mechanical nature acquired by manual practice, and those of a scientific character, to be learned only by study in connection with that practice. The phenomena which occur in the daily routine of the apprentice, though matter for wonder at the beginning, become familiar by repetition, cease to excite thought or inquiry, and he rests satisfied in ignorance of their nature. But when, during his practical lessons, a course of study is pointed out to him, he soon gets a key to much that was hidden, becomes interested in what he is doing, and progresses rapidly in proportion. It often happens that the beginner does not enjoy the privilege of a friend at his side, to explain difficulties as they arise, and employers are sometimes as ignorant in these regards as the apprentice himself. It is to point out to these seekers after knowledge the route to obtain it, that this Syllabus has been prepared.[73]

Ironically, several years later Albert Prescott used some of Procter's arguments against Procter and his stand on the primacy of apprenticeship in pharmacy education.

Upon publication the "Syllabus" received good reviews, for it offered an excellent guide for those students who could not attend formal lectures.[74] In his review, Edward Parrish recognized the syllabus for what it was: a transcription of Procter's course in practical pharmacy at the College, which was based on an arrangement by the French pharmacist-chemist Soubeiran. To the pragmatic Parrish, the syllabus seemed more "scientific than convenient, though perhaps the best for a student who would aim at a thorough knowledge of theory and practice of pharmacy."[75] Throughout the rest of the nineteenth century Procter's "Syllabus" influenced the way American pharmacy was taught in the classroom, in correspondence courses, and in pharmacies by the preceptors of apprentices. It served as an authoritative alternative to the practice of merely going through the *Pharmacopoeia* or *Dispensatory* from A to Z. Through this syllabus Procter guided the education of thousands of American pharmacists.[76]

During the decades that Procter taught at the College, he never abandoned the basic framework of his syllabus even though American pharmacy went through profound changes. He refused to allow these trends to alter the basic subject matter of his course.[77] Procter tended to pursue his concept of ideal pharmacy, often ignoring the reality of drugstore practice.[78] Unfortunately, this conservative and rigid approach to the teaching of pharmacy continued to be the rule for the next hundred years.

When the young pharmaceutical industry grew swiftly during the late 1850s, Procter noticed the trend and reacted: "Pharmacy may be defined to be the art of preparing and dispensing medicines, and embodies the knowledge and skill requisite to carry them out in practice. But if the preparation of medicines is taken from the apothecary and he become merely the dispenser of them, his business is shorn of half its dignity and importance, and he relapses into a simple shopkeeper."[79] For this reason Procter insisted that his students adhere to his ideal and learn how to make all their own medicines. Even after it became evident that most pharmacists were turning manufacturing over to others, he continued to support the primacy of apprenticeship over school learning.[80]

In the years after the Civil War, some physicians began calling for laws that would regulate the practice of pharmacy. At first Procter agreed that some sort of educational requirement might be the answer.[81] Yet, when the Philadelphia County Medical Society drew up a law that would require pharmacists to receive a pharmacy school diploma before they could open a shop, Procter objected. He committed himself to the perpetuation of the apprenticeship system, even though the proposed law would have greatly enlarged his classes and his income. He strongly supported pharmacy school as a "rounding off" of the young pharmacist, but could not give formal education priority over practice experience. He feared that the law proposed by the medical society would close half the shops in Philadelphia, thereby eliminating apprenticeship opportunities.

Why not just teach pharmacy in the classroom and forgo shop experience? To Procter, formal education without the preparatory apprenticeship was "like requiring a child to learn to swim without going into the water, and strikes at the very root of progress in skill and knowledge."[82] His view dominated the policies of the College of Pharmacy, even after he retired.[83]

The changing pharmaceutical scene of the mid-nineteenth century concerned other pharmaceutical educators more than Procter. Edward Parrish, who had run a practical school of pharmacy since 1849, led efforts during the 1850s and 1860s to have the College establish a laboratory for practical pharmacy training.[84] Procter supported the idea early, but not vigorously.[85] He was content to allow Parrish to continue to teach those prospective pharmacists who could not get adequate apprenticeship experiences.[86]

In 1865 the new Alumni Association of the Philadelphia College of Pharmacy proposed the creation of a practical laboratory in pharmacy to be associated with the College. After initially reacting favorably to the idea, Procter expressed some serious doubts. He feared that the prime movers behind the practical school might be more interested in collecting fees than educating students. To Procter the single most important question was:

whether this instruction is intended to *perfect* students who may be in shops unfavorable for gaining a practical experience, or whether it is to be a substitute for the shop? [I] hope not the latter. While the curriculum should involve an examination in the rudiments of Pharmacy, and where these have been but imperfectly attained it should insist on their acquirement, yet the main and true object of the School should be to give perfect lessons in pharmaceutical manipulation with the best forms of apparatus, and to afford the student full opportunities to become skilful in their uses and applications, especially of those which the tendency of latter years had nearly banished from the shop.[87]

Procter did not object to laboratory instruction per se, but to its substitution for in-shop experience.[88]

As work on the new laboratory progressed Procter wrote to his friend Albert Ebert: "Our Practical School is a subject of great anxiety to me & I regret much not having given the time and attention when in Europe . . . to learn the real manner of working the tuition of practical pharmacy. The trouble here is to find the right man for the place. It must be someone who can give the time 5 days in 7 to it from 9 AM to 3 PM in the year & then cover the practical difficulties of keeping order & industrious habits among the class."[89]

The Alumni Association completed the laboratory in 1870 and placed it under the able direction of John Maisch, professor of materia medica at the College and a skilled pharmaceutical chemist. In addition to teaching pharmacy, Maisch offered a separate course in analytical chemistry, which became more popular probably because of the continuing problem of adulterated drugs and chemicals. Laboratory space for twenty students soon filled up, and in 1872 twelve additional tables were added. Because of the success of the laboratory, it became too large a burden for the Alumni Association, which turned it over to the College in 1872. For all of his encouragement of the practical laboratory, Procter never advocated it as a preferable substitute for apprenticeship experience.[90]

The expansion of pharmaceutical education into state institutions of higher learning stimulated discussion of the benefits of apprenticeship versus formal education. As editor of the *American*

Journal of Pharmacy Procter had observed the slow growth of pharmaceutical education across the United States. In 1867 he gave notice to the new pharmacy course at the University of Michigan, calling it "very useful to those who attend."[91] Two years later, however, Procter grew skeptical about the value of the Michigan course. It had come to his attention that this program did not require practical experience for graduation.[92] He observed that an "apothecary without shop experience is like a medical graduate without hospital or other practice: they are both of doubtful reliability."[93] Albert B. Prescott, a physician who taught pharmacy at Michigan, responded to Procter's criticisms: "Our school believes it to be quite as well for the young pharmacist, better for his employer, and far better for the public that scientific preparation for the drug business should *precede* experience in it."[94] To this Procter retorted:

This is all well enough as regards the preparation of the student for his pharmaceutical duties, but to give a diploma to a student intimating that he is a pharmaceutical chemist, which means an apothecary or pharmaceutist, when he has no practical familiarity with drugs and with shop experience, is not right. The point of the matter is whether the latter class leave the University and offer themselves as qualified clerks, or whether they enter as beginners? In any case the school authorities should adopt some other title for their diploma given to such, and not that which in the most thoroughly practical English pharmaceutical school is given by act of Parliament only to accomplished pharmaceutists.[95]

Michigan's new approach to pharmaceutical education served as the focus of the first meeting of the Conference of Teaching Colleges of Pharmacy in 1870.[96] At the conference, delegates met from seven pharmaceutical societies that taught pharmacy. The issues discussed included whether standards should be established, the length of apprenticeships, what subjects should be taught, what texts should be used, and should apprenticeship be required for a diploma.[97] Using the issue of preliminary education of apprentices as a springboard, Procter successfully offered a motion that diplomas would not be recognized by the conference unless the student had served for four years as an apprentice in a dispensing shop.[98] The conference adopted this position, and the

Maryland College of Pharmacy at their next meeting added this requirement to their charter.[99]

Procter felt hard pressed to evaluate the relative value of formal education against traditional apprenticeship and so wrote an article entitled "Can Practical Pharmacy Be Taught Effectively by Lectures?" After praising the constructive influence of classroom instruction, he took this equivocal stance: "All will agree that no amount of tuition by lectures will be equivalent to that which the earnest student receives in the dispensing shop and practical laboratory, under the personal instruction of a well qualified pharmaceutist, who takes an interest in his pupil; yet such opportunities are rare."[100] Unfortunately, Procter never answered his rhetorical question straight away; instead he merely advised his readers on how to make lecture courses more interesting through demonstrations.[101]

Procter realized how poorly most apprentices were prepared, but saw no good alternative to apprenticeship. He could not conceive a course in practical pharmacy that would suffice. Even after the success of the Alumni Association laboratory, Procter did not foresee how the independent schools, usually penurious, could provide such facilities for a full class of students. Thus he continued to support the apprenticeship system and hoped that it would improve. Other pharmaceutical educators, such as Prescott, who accepted the state of pharmaceutical practice as it was in the late nineteenth century, felt confident that classroom and laboratory instruction could prepare students for drugstore work.

The conflict between the traditional apprenticeship system and the new university approach came to a head at the 1871 meeting of the American Pharmaceutical Association held in St. Louis. Prescott traveled down from Ann Arbor as a delegate to the convention. Several members of the association, led by its permanent secretary John M. Maisch, objected to the seating of Prescott as delegate because "the Michigan school is not what [the] By-laws regard as a college of pharmacy or local pharmaceutical association."[102] The discussion soon shifted from the status of the Michigan school as a pharmaceutical organization to the value of its pharmacy course. The apparent presumptuousness of the Michigan school to christen inexperienced youths as pharmacists outraged several members. Some suggested that Dr. Prescott be asked to leave the convention because he was a medical school graduate and never practiced pharmacy.[103]

Parrish, always trying to keep the association as open as possible, remarked:

Pharmaceutical education, in this country, is in its infancy. There are very few schools that teach Pharmacy; there are very few persons who are graduates in Pharmacy. We have heard it said that a graduate of medical school has no right among us. Let us remember that there are many here who have never even graduated in a school of pharmacy, and who make no pretensions to it; but we give them the right hand of fellowship, and we want them to be associated with us for their good and our good. We are comprehensive. We are not a college elected after a great care and research as to the qualifications of each individual; we take everybody in, and if the delegate from the University of Michigan came here as an individual, there would not be the slightest question about his admission.[104]

Parrish proposed successfully that a committee made up of a representative from each of the organizations at the convention should decide the question.[105]

The report of this committee reveals the attitude of the old line:

The Committee . . . are united in the conclusion that the University [of Michigan] is not, within the proper meaning of our Constitution and By-laws, a College of Pharmacy: it being neither an organization controlled by pharmacists, nor an institution of learning which, by its rules and requirements, insures to its graduates the proper practical training, to place them on a par with the graduates of the several colleges of pharmacy represented in this Association. We therefore recommend that the credentials of the delegate from the University of Michigan, be returned to him with a copy of this report.[106]

The body of the convention accepted the report "without a dissenting vote."[107]

This action did not end the discussion. Prescott, accepted as an individual member of the association, quickly composed a paper on pharmaceutical education and read it during the session for volunteer reports. With the paper's opening line, Prescott fired the first salvo in a war of words that would last several years: "The conditions of pharmaceutical apprenticeship in this country constitute a subject of regret to all thoughtful observers." Prescott made the radical suggestion that school learning should precede shop experience, arguing that this arrangement would make the apprenticeship more worthwhile. Moreover, by working in a pharmaceutical laboratory, students learn far more than through lectures alone.[108]

John M. Maisch, secretary of the association and a professor at the Philadelphia College of Pharmacy, reacted strongly to Prescott's views, calling them "entirely erroneous." He praised Prescott's advocacy of the basic sciences of pharmacy students, but insisted that "with all his knowledge of chemistry, natural history, and natural sciences generally, he will not be a pharmacist until he has gone through a regular system of training, and that is exactly where the colleges of pharmacy throughout the country differ from the University of Michigan."[109] Although the proceedings recorded no comment by Procter on this matter, his views coincided with those of Maisch. Thus Procter and other pharmaceutical educators missed an opportunity to place pharmacy more on a par with the medical profession through embracing the primacy of formal education.[110]

During the decades that Procter taught pharmacy, he witnessed pharmaceutical education grow from two active schools, those in New York and Philadelphia, to about half a dozen of varying vitality. Except for the University of Michigan course, these were all progeny of local pharmaceutical societies. As editor of the *American Journal of Pharmacy,* Procter made the encouragement of pharmacy schools one of his personal missions.[111] He viewed pharmacy schools as the natural outgrowth of the spread of local pharmaceutical societies across the country. When these local groups had difficulties, he often came to their rescue, providing needed help and advice.[112]

In summary, William Procter aided the growth and progress of pharmaceutical education in the United States through his precedent-setting tenure as professor of pharmacy, his textbook *Practical Pharmacy,* his "Syllabus," and perhaps most of all, by educating many of the next generation of professional and educational leaders. By showing that practical pharmacy had a base in science, Procter deserves credit for starting the academic discipline of pharmaceutical chemistry in the United States. Yet, for all his accomplishments in the field of pharmaceutical education, Procter and his authoritative voice played a part in American pharmacy's failure to attain in full the sort of professional position and status gained through high educational standards. Because he championed the primacy of practical experience over book learning, Procter missed an opportunity to influence significantly the reform of pharmaceutical education. In addition, he chose to ignore the phenomenon of drug and medicine manufacture shifting from the shop to the factory. He set a precedent that pharmacists should learn how to make every dosage form that they dispensed—a viewpoint maintained for the next hundred years.[113]

Procter foresaw the increased influence of the state on pharmacy, but largely rejected the positive effects regulation could have for his profession. Educational requirements for practice imposed by the state clashed with Procter's commitment to intraprofessional control. For Procter, the state could test for competency, but could not expect all pharmacists to earn a diploma before practicing.

As a teacher of pharmacy, Procter should be seen as a transitional figure between traditional pharmacy practices and the professional posture and status American pharmacists achieved by the end of the nineteenth century through the adoption of modern, credential-based professionalism. A conservative, Procter tried to maintain the best aspects of older practices, tempered and adjusted to fit the difficulties presented by changes of the mid-nineteenth century. He demonstrated through his own example that pharmacists could teach their own craft scientifically. During his tenure as professor of pharmacy Procter set the model for association schools to follow: The professor of pharmacy should convey the basics of practical knowledge and technique to students so that they could competently practice pharmacy (or pass a state board examination).[114]

Moreover, Procter contributed to the evolution of pharmaceutical education in the United States by training a large segment of the next generation of teachers. One reason practical pharmacy attained its position as a discipline so slowly was the paucity of competent instructors willing to work for low compensation.[115] Parrish went so far as to claim that pharmacy's subservient relationship to the medical profession could be attributed to the fact that physicians dominated pharmaceutical education.[116]

Several of Procter's students helped reform pharmaceutical education in the late nineteenth century. Frederick B. Power, as director of the pharmacy course at the University of Wisconsin, extended Procter's ideal of the practitioner-scientist to include students. Instead of just practicing shop manipulations, Power's students did research, which the department collated and published.[117] Henry S. Wellcome did not become a teacher of pharmacy, but improved the quality of pharmaceutical education through his philanthropy.[118] And Joseph P. Remington succeeded Procter at the Philadelphia College of Pharmacy, where he taught for the remarkable tenure of forty-four years and wrote the most popular textbook in the history of American pharmacy.[119] All three men listened to Procter deliver his last lecture just hours before he died.[120]

Procter and Pharmaceutical Organizations

7

At the first general session of the 1874 annual meeting of the American Pharmaceutical Association, President John Hancock turned from routine business to remark on the recent death of Procter: "Instinctively I turn my eyes with a sort of vacant stare to the seats in your midst, as if to recognize his presence; his voice, so familiar, seems to fall upon my ear. But we will never be encouraged by his presence any more. Let us imitate his example."[1] What sort of man inspired Hancock and others to express themselves in this fashion?[2] An examination of Procter's organizational work documents his leadership, but even more it reveals the consummate association member.

Procter is rightly remembered as one of the founders of the American Pharmaceutical Association. Urdang went so far as to say that the founding "was mainly due to his foresight and persistence."[3] Though the association was "the favorite child of his genius," his organizational training came early in life through his relationship with the Philadelphia College of Pharmacy.[4] Procter came to believe that such local organizations were on the forefront of pharmaceutical progress and therefore put their encouragement high on his list of professional missions:

[T]here is no reason why every city and large town should not have a local Society of Pharmacy, organized with a view to the professional, scientific and social advancement of its members. "In union there is strength" is a political motto equally applicable to our profession. Under the fostering influence of such local unions, libraries of scientific books may grow up, the best journals of scientific literature in all collateral branches can be subscribed to, forming a central reading room of science, where the members may meet for self-improvement or discussion. Under such a union the "rough corners of trade jealousy" are rubbed smooth, pharmaceutists are willing to act together for mutual protection or advancement, and the evils of an unhealthy and unrighteous competition are at least abated and controlled.[5]

Procter always insisted that the membership of such organizations should remain open to "all of good intention." Commenting in 1856 on the rebirth of the defunct Maryland College of Pharmacy, he wrote, "The object of such organizations should be to raise the status of the entire body; its fruits should not be for those already enlightened, so much as for those who desire improvement; and the prominent part taken by the more educated should be viewed, as it generally deserves to be, as their good will offering. We are far from advocating an indiscriminate admission, but every one, however small his pretensions, who has correct views of the vocation of the apothecary, should be included."[6]

For Procter, one of those "correct views" was the ready acceptance of the duty to join a local association. In his opinion, the cause was clear: to make each pharmacist a competent compounder of medicines and an ethical businessman. He argued that local organizations could do this through educational programs and efforts to reduce destructive competition. Procter did not favor fee schedules, but thought that a general agreement to compete on the quality of service instead of price would improve the general level of practice. Because this could only occur when most pharmacists participated, Procter favored open organizations.[7]

Procter's view on the purpose of such groups was not typical. His early career had been rather unusual. He had trained in one of the leading shops in Philadelphia, but had to open a modest establishment because of lack of capital. He may have seen himself as living proof of the benefits of College affiliation—a poor boy who could aspire to the calling of Scheele, Pelletier, and Caventou. Yet in the early and mid-nineteenth century, pharmaceutical societies like the Philadelphia College of Pharmacy were made up mainly of well-established pharmacists, many of whom had large wholesale businesses. Several of these early groups had arisen in response to the possibility of outside control of pharmaceutical practices by medical societies. The resulting organizations were usually closely knit and conservative. Because they tended toward elitism they failed to attract the majority of practitioners needed to institute some measure of control over local practices, such as fee schedules, the character of apprenticeships, and shop hours. Instead, members of pharmaceutical societies focused on self-improvement and elevating their status within the community at large.

Aware of these shortcomings, Procter still saw great potential in pharmacy organizations. As editor of the *American Journal of Pharmacy* he included announcements concerning various local societies in almost every issue. This action might be partly attributed to his efforts to make the *Journal* a national organ, but his enthusiastic editorials went beyond mere journalism. From the beginning to the end of his editorial tenure, Procter praised the efforts of local groups, giving them needed encouragement and advice.[8]

Just as Procter printed the minutes of the various local pharmaceutical organizations across the country, he also announced their journalistic efforts. Apparently Procter did not fear competition from other college journals, like those of the New York and Maryland colleges. Leaving the business side of pharmacy to the *Druggists' Circular,* Procter hoped that the college journals would stimulate scientific investigation and serve as conduits for pharmaceutical education.[9]

In addition to being a strong advocate of local pharmaceutical organizations, Procter actively participated in the Philadelphia College of Pharmacy. Three years after his graduation in 1840, he was elected a resident member of the College, together with his equally talented friend Augustine Duhamel.[10] In his first year of membership, Procter only occasionally attended business meetings of the College, but after his election as a trustee (1841), he became a fixture at College events.[11]

Over the next thirty years, Procter gave of himself to the College in many ways: in money, books, and specimens, but most of all in time. He served on numerous committees, which considered issues from the mundane to the weighty. For example, in 1862 he helped collect materia medica specimens for the World's Fair in London, only to be prevented from sending them because of the Trent affair.[12] In 1872 the College appointed Procter chairman of a committee to obtain a complete set of the *American Journal of Pharmacy* and other books to send to the library of the Chicago College of Pharmacy to replace those lost in the Chicago fire.[13]

Procter also held various elected offices during his thirty-five-year connection with the College. As a trustee, he helped supervise the school of pharmacy through the appointment of professors and the committee of examination. In addition, the trustees administered the election of members and the physical plant of the College.[14] In 1855 Procter's responsibilities expanded with his election as corresponding secretary of the College. As editor of the *Journal* he had been performing some of the duties of this office anyway. He held the position from 1855 to 1867, which were challenging times for the College. In 1867 Procter became first vice-president of the College, with Alfred B. Taylor taking over the job of corresponding secretary.[15] Procter held this primarily honorary position until his death, except for the year of his European trip.[16]

In his dedication to the College he was exceptional, but not unique. A small cadre of members kept the College ahead of all other local pharmaceutical societies in the country by supporting the *Journal* and the lecture courses. Some like Daniel B. Smith and Dillwyn Parrish worked for the College as part of a larger range of civic responsibilities; others such as Charles Ellis and Joseph Turnpenny had a special devotion to the cause of pharmacy.

If Procter had not been there in 1846 to teach practical pharmacy, Edward Parrish could have filled the position (as he did twenty years later); the same can be said of the editorship of the *Journal*.[17] Parrish was available and capable. The College would have grown and prospered without William Procter, Jr. Yet his presence did make a difference. Procter possessed a mixture of pharmaceutical skills and personal character that carried the College to heights probably unattainable without him. He combined the practical interests and expertise of Parrish with a respect for science and research absent in the younger pharmacist. Moreover, as editor he provided a firm, even-tempered voice that served the College better than the verbosity and polemic of Parrish's prose. Above all, Procter served as a steadying force. Not an innovator comparable to Parrish, he was a preserver of past achievements and an advocate of progress. This can be seen in his early championing of the American Pharmaceutical Association, which was, in essence, a national extension of what he perceived to be the best attribute of the Philadelphia College of Pharmacy: a tool for raising the competency and scientific knowledge of the common practitioner of pharmacy.

The American Pharmaceutical Association (APhA) was founded in Philadelphia in 1852, but for its beginnings one must go to Washington, D.C., in 1848. On June 26 of that year, President James Polk signed an act "to prevent the importation of adulterated and spurious drugs and medicines."[18] Soon afterward, inspectors set to work at six main ports (New York, Boston, Philadelphia, Baltimore, Charleston, and New Orleans) and began refusing entry to drugs they judged inferior. The reputable pharmacists of New York, represented by their college of pharmacy, objected to what they saw as laxity on the part of their inspector. (He had been unable to reject some of the worst lots of drugs because recognized objective standards, based on percentage of active constituent, did not exist.) The New York pharmacists had been especially active in the effort to get the law passed and wanted stiffer enforcement. Efforts to work through the young American Medical Association and other groups had failed, so the New York College of Pharmacy invited all other pharmacy societies to come to a convention to discuss the problem in October 1851.[19]

The invitation elicited a positive response among concerned pharmacists in Philadelphia. As Procter recalled in a later editorial:

When the invitation to the Convention . . . was received by the Philadelphia College of Pharmacy, several of the members expressed the opinion, that although the call was for a special object, the Convention might take a wider range in its influence, and form a *point d'appui* from which the pharmaceutical profession of the whole country might be reached, and a course of action instituted, which eventually would revolutionize the condition of Pharmacy in the United States.[20]

Procter, an admirer of the national character of the Pharmaceutical Society of Great Britain (f. 1841), hoped that a similar model organization would be created in the United States.[21] Together with fellow delegates Charles Ellis and Alfred Taylor, he traveled to New York with these aspirations in mind.

On October 15, 1851, at 5 P.M., the delegates called to New York met at 511 Broadway to begin their discussion of the drug importation law and any other "measures that might be calculated to elevate the profession, and promote their interests throught the country."[22] Only nine delegates attended, three each from the colleges of New York, Philadelphia, and Boston; no delegates from Cincinnati or Baltimore arrived. After the preliminaries, a committee of Procter, George Coggeshall of New York, and Thomas Restieaux of Boston was appointed to draw up the business of the convention, which then adjourned for the night.

Procter and his committee worked long into the night on a report that addressed the principal issue of the convention, standards for imported drugs, and the additional subject of a national organization. The next morning, Procter read the report to the small group of concerned men, where it "was considered in sections with full, deliberate, and very interesting discussion."[23]

After agreeing to a new set of standards for imported drugs, the convention participants assessed the value of a national organization. The assembled group "considered and adopted" the following preamble and resolutions read by Procter:

Whereas, The advancement of the true interests of the great body of pharmaceutical practitioners in all sections of our country is a subject worthy of earnest consideration; and *whereas,* pharmaceutists, in their intercourse among themselves, with physicians and the public, should be governed by a code of ethics calculated to elevate the standard and improve the practice of their art; and *whereas,* it is greatly to be desired that the united action of the profession should be directed to the accomplishment of these objects; there, *Resolved,* That in the opinion of this convention much good will result from a more extended intercourse between the pharmaceutists of the several sections of the Union, by which their customs and practice may be assimilated; that pharmaceutists would promote their individual interests, and advance their professional standing, by forming associations for mutual protection, and the education of their assistants when such associations have become sufficiently matured; and that, in view of these important ends, it is further *Resolved,* That a convention be called, consisting of three delegates each, from incorporated and unincorporated pharmaceutical societies, to meet in Philadelphia on the first Wednesday in October, 1852, when all the important questions bearing on the profession may be considered, and measures adopted for the organization of a National Association, to meet every year.[24]

The convention then appointed Procter, Coggeshall, and Samuel Colcord of Boston as a committee to organize the meeting.[25]

Thus the American Pharmaceutical Association was conceived in the words of Procter. He had succeeded in getting agreement on what concerned him most: that pharmacists needed to follow a code of ethics and strive for some uniform standard of practice for the calling to gain public respect. Once back in Philadelphia, Procter set to work to stimulate interest in the upcoming meeting through his editorials in the *American Journal of Pharmacy.*

During the early months of 1852, Procter contemplated the upcoming meeting and its agenda. Besides the obvious mechanics of setting up a new organization, he wanted the association to address the problem of improving the level of practice among American pharmacists above that of a mere trade. In an editorial, Procter posed a number of questions for the convention to consider:

1. By what means can the existing large number of ill qualified practicing pharmaceutists, everywhere over our country, be induced to improve themselves in education . . . without omitting their present duties? 2. What are the most efficient and best adapted means by which our apprentices . . . can receive the benefits of pharmaceutical education? 3. How far will the principle of association enable the better qualified to extend assistance to the deficient? 4. What means are best calculated to sever the existing connection between pharmacy and quackery, and to induce apothecaries to repudiate the sale of secret remedies? and lastly, what suggestions can be offered to the convention, by which it may hold out

inducements sufficient to engage and direct the latent talent of our ranks, to such useful and interesting scientific objects as shall redound to the improvement of our profession at home, and its reputation abroad?[26]

Such editorials stimulated interest in the convention among the *Journal*'s readership. Pharmacists who read the *Journal* but were not members of an association wrote to Procter asking him how they could participate, since this information had not been in the original call to the convention. Procter took it on himself to suggest "that any *ten* established apothecaries and druggists, located in one place or neighborhood, where *no organization exists,* should be entitled to send a delegate to the Convention." Small towns, with less than ten apothecaries, could send a representative. Other correspondents asked Procter what the convention would do: just pass resolutions and go home? No, said Procter, "It is action, not talking, that is wanted; it is wisdom, not eloquence, that is needed; it is conscientious pursuit of the general good, not individual ambition, that is desired; and it is a generous sacrifice of knowledge and influence by the better educated and successful, on behalf of those whose want of qualifications and success arises more from the force of circumstances, than from disinclination to improve their practice."[27] And again Procter delineated his ideas for the first agenda for this association that had never met. The best plan for local organization headed the list or priorities, with pharmaceutical education and the stimulation of research not far behind.

As with many of his editorials from this period, Procter finished with a little hortatory prose: "Deep rooted evils in a profession can rarely be removed by sweeping legislation, unless the measures are enforced by despotic power. Let the well disposed among us, therefore, show practically the working of a higher standard, as an example to those not now willing or able to adopt it, which will be more influential than volumes of precepts."[28]

As chairman of the executive committee of that first APhA convention, Procter successfully argued that the organizational questions had to be handled first. He feared that controversial issues such as a code of ethics or the reform of pharmaceutical education might consume too much of the convention's time.[29]

Before anything could be decided, however, the matter of membership requirements had to be settled. Procter moved that all interested pharmaceutists and druggists could participate; others, though, wanted to restrict the association to those who already belonged to an organized pharmaceutical society. They viewed the association as a sort of honorary body, which would add to their personal prestige. Procter's position received support from his eloquent, and often quick-tempered, friend Edward Parrish, who contended that anyone in the trade who possessed a character "morally and professionally . . . fair" should be admitted. After an "animated debate" the "liberal view" was adopted.[30] In this way the young APhA avoided the exclusiveness that led to the demise of the first national dental organization.[31]

The first APhA meeting satisfied Procter: the problems with the drug law had been discussed, along with the sale of poisons and the subject of secret remedies. Above all, the convention resolved its organizational policies and directly addressed educational issues, two priorities of Procter's for the meeting.[32]

Procter would live to attend twenty more meetings of the association. For the first five years of the APhA's existence, Procter served as corresponding secretary, the organization's only continuing officer. Procter was the obvious choice for this position because he had been doing similar work informally as editor of the *American Journal of Pharmacy,* the only national pharmaceutical journal of the time. In the early years of the association, Procter also served on the executive committee. At the 1853 meeting, for instance, Procter as chairman of the executive committee read a report (written with the assistance of George Coggeshall, though the prose is Procter's) that fulfilled the duties given to the committee at the convention in 1852. These included the first detailed report on the status of pharmacy practice in the United States and a code of ethics later adopted by the association.[33]

Over the years the membership repeatedly selected Procter for positions of authority. In 1859 he was elected first vice-president of the APhA (after refusing nomination the year before). In 1862 he was elected president. Aside from running the meetings and giving an address, the presidency was honorific. Procter exerted his leadership in the young association mainly through his work as a member, through committee service rather than as an officer. Procter endeavored to attend every meeting, even in distant cities such as Cincinnati, Detroit, and Chicago. His example as a member was one of dedication and humility. Samuel Colcord, a rather brusque, Boston pharmacist, once chastised Procter for his "excessive modesty" in organizational matters.[34]

In the early years when the APhA had a small number of members (e.g., seventy-five in 1855), Procter served on most committees. This occurred not only because Procter was a willing and good worker, but because he often had suggested the creation of the committee in the first place. Other appointments came to him because of his acknowledged expertise as pharmaceutist, editor, or author. In 1855, for instance, he was appointed chairman of a committee to select the best textbooks for pharmacy study and was a member of five of the other seven standing committees of the association.[35]

Procter exhibited his pharmaceutical knowledge through his chairmanship of the committee on queries. In an editorial written before the founding meeting, he suggested that "qualified individuals, might be entrusted with . . . special subjects for investigation" and gave six possibilities. When no one seemed interested in taking on these challenges Procter changed his tactics. In 1854 he called for the establishment of a queries committee, based on the model of the French Academy of Science and volunteered his own service.[36] The association agreed and Procter set to work as chairman of the new committee, thinking up practical and scientific problems. At each annual meeting volunteers would pledge to answer certain "queries" at the subsequent meeting.

Because of his devotion to the ideal of the association as a forum for scientific discussion, Procter took on, year after year, the difficult task of thinking up questions for investigation and getting members to accept them. In 1867 Edward Parrish remarked that "[f]rom my knowledge of Prof. Procter and the Association, I think he is the only man in it that can get up these queries. It is always a source of amazement to me how he gets hold of them."[37]

Procter composed scientific queries for members in one capacity or another until 1873.[38]

Over the years of his involvement in the association, Procter never retreated from his belief that no feature of the organization was more important than the stimulation, reading, and publishing of scientific reports on pharmacy.[39] Judging by the number of scientific papers published in the annual proceedings, compared to the number published previously in the *American Journal of Pharmacy,* Procter's advocacy of association-sponsored queries roughly doubled the amount of scientific investigations undertaken by American pharmacists. A few examples of the better papers presented at association meetings that came out of such queries included: Israel J. Grahame on "The Process of Percolation or Displacement,"[40] Charles T. Carney on "Paraffin—Its Substitution for Wax in Cerates,"[41] and Procter's own "Remarks on Dialysis."[42]

Procter stimulated general participation in the association through his writings in the *Journal.* Especially in the early years of the association, when the *Journal* provided the only national medium for pharmaceutical news and discussion, Procter all but set the agendas for upcoming meetings through his editorials. From the beginning in 1852, Procter decided to publish the proceedings of the business meetings in the *Journal,* which brought them to the attention of a wider audience.[43]

In his editorials concerning association meetings, Procter sought to attract more pharmaceutists to them, preferably with scientific papers in hand. His message changed over the years, but usually included predictions of the upcoming agendas (more than educated guesses, for he often served on the executive committee or as an officer) and the sights to see in the host city. After the meeting Procter's editorials would review the best papers given, the key debates, and more and more as years went by, the festivities that accompanied the meetings.[44]

In addition to calling for better attendance at the annual meetings, Procter's editorials advocated an adherence to the original purpose of the association through the scientific and ethical improvement of the individual pharmaceutist. Procter stressed the relative unity of at least a group-conscious minority of American pharmacists, which he contrasted with the disunity of the medical profession generated by sectarian bickering. At association meetings, in Procter's view, questions of science and art should transcend those of commercial competition, sectarianism, and sectionalism. Throughout the 1850s, as the nation's "impending crisis" flickered on the horizon, the association followed Procter's lead, avoiding regional issues and sticking either to practice concerns or national problems such as faulty implementation of the drug importation law. As the decade of the 1860s began, however, this national approach was severely tested.

Trying to do its own part to foster the good health of the Union, the association decided in 1860 to hold its next meeting in St. Louis. Despite talk of war following the election of Lincoln, Procter hoped that the membership would not "lose sight of the importance of this benevolent and disinterested scientific movement in favor of the improvement of pharmacy and pharmaceutists—a movement so catholic in its character that it should proceed uninfluenced by the political aspects of the country." He looked forward to a St. Louis meeting that would bring a renewed interest in the association among Southern members.[45]

Even after the shelling of Fort Sumter, Procter held out the hope that the St. Louis meeting would be held:

In view of the disordered condition of public affairs, it is doubtful whether there will be a large attendance even from places adjacent to the place of meeting, and it is all the more important that the reports should go on, so that at least some interest may be obtained for the Proceedings of 1861. As a *scientific* body, national and even continental in its scope, the American Pharmaceutical Association should live on through all the changes of a political character, and hence, whatever else be done, it is of vital importance that the regular preliminary forms for the meeting should be carried out, and that a meeting should occur, if but a dozen members be present. . . . [I]t is to be hoped that ere the period of convening, the dark clouds which now lower over our beloved country will be dispersed.[46]

This was not to be, and like the American Medical Association (AMA), the APhA did not meet in 1861.[47] Procter tried to rally members by encouraging them to carry on with their assigned queries or voluntary papers. Two years of work should produce much good, and "should the progress of events favor the settlement of our national difficulties, so that the currents of our business and social relations may return to the wanted channels, we may look for many valuable contributions to our literature, not to speak of the pleasures of friendly intercourse renewed after so long a separation."[48]

Instead of deferring the 1862 annual meeting, as the AMA did, Procter insisted that it be held in Philadelphia to commemorate the tenth anniversary of the founding. He did not write of war's end; instead he stressed the importance of finishing papers and getting them into print. If the situation did not allow members to come, they could always send along their papers with the few who planned to attend.[49] Even with Procter's urgings, the 1862 meeting was poorly attended, mostly by members from the host city, Philadelphia, and nearby New York.[50] At that 1862 annual meeting Procter was elected president of the association, a fitting reward for the man whose voice had played a large part in keeping the young organization alive during the hardest days of the Civil War.

At the war continued, Procter also continued to reassure the members of the association through his editorials in the *American Journal of Pharmacy*: "We know how prone men are to be attracted from their ordinary duties by the occurrence of such events as are now happening near us, yet while prepared to do our duty as citizens, we should not forget that our duty as pharmaceutists is to sustain our National Association."[51]

When the war ended Procter did not praise the gallant armies of the Union and their prowess in battle; he welcomed back the southern members of the association and reminded them of their unfulfilled assignments to answer queries. He returned to writing editorials that urged pharmaceutists to join the association, to do investigations, and to give papers at the annual meetings. He praised those who gave excellent papers and reserved his harsh words for the poor printing of the postwar proceedings.[52] Finally, in his last few years as editor, Procter became more critical of those members of the association who put mammon before pharmaceutical art and science.

During Procter's twenty-two year relationship with the American Pharmaceutical Association, he participated in numerous debates concerning a variety of issues.[53] By comparing Procter's attitudes on certain issues with those of other members of the association, his role as leader and member can be gauged.

When debate raged at an association meeting Procter usually hung back, listening to arguments and mulling over both sides of the controversy. He then would get the floor and summarize the debate as succinctly as possible, often boiling down the discussion to a simple choice among a few alternatives. Members repeatedly expressed gratitude to Procter for this ability and often appointed him to committees that were set up to study controversial issues.[54] When Procter did hold strong opinions himself, however, he readily expressed them. His reactive nature let others create a controversy by their words or deeds, to which he then responded with a sense of principle and mission.

An early debate centered on who should be invited to join the new society. From the very beginning Procter and his friend Edward Parrish worked to keep the association open to any interested pharmacist.[55] When the association decided to admit any pharmacist who agreed to follow its code of ethics, Parrish expressed concern that a strict interpretation of the obligation "to discountenance quackery" would force out many potential members. He pointed out that it was "mainly by the sale of quack medicines that many druggists subsist, who yet desire a reform in their business."[56] Procter, who had written much of the code, thought that the association should ignore the issue for the time being. A realist in this matter, he differentiated between selling nostrums and recommending their use. He himself did not stock them, but did not chastise pharmacists who did. In a footnote to a notice of the 1855 annual meeting, Procter said that it "should be distinctly understood that all reputable Pharmaceutists and druggists, who are favorable to the objects of the Association will be welcomed at the meeting."[57]

Procter's democratic attitude toward the association is well illustrated by his comments concerning the newly formed British Pharmaceutical Conference (1863), which was modeled largely after the American Pharmaceutical Association. He hoped that the new British group would "tend to elevate the status of that numerous class now known as chemists and druggists. To be successful, the leading spirits will have to guard it from a tendency to learned exclusivism and titular pride; its action must be as far as possible democratic, and based on the broad principle of encouragement to merit in every class with the unremitting object of improving the actual practice of Pharmacy in every shire and town in the British Islands."[58]

In 1868 the association, primarily on the urging of then President Parrish, expunged the code of ethics from its official documents. Procter, chairman of a committee appointed to evaluate the code, reported that inasmuch as new members signed the constitution of the association, which he felt contained the essential ethical material, the code was unnecessary. As a principal author of the code, Procter probably convinced the membership to drop it.[59]

Ironically, the first real test of the association's ethical standards came the next year. Long-time member Frederick Stearns, a drug manufacturer from Detroit, had begun to market a product called Sweet Quinine. Several members were outraged when they discovered that the product did not contain quinine but the related alkaloid cinchonine. Stearns contended that there was no therapeutic difference between the two cinchona alkaloids and that his new preparation had distinctive advantages over others made from crude cinchona bark. Above all, Sweet Quinine contained a fixed amount of alkaloid, in contrast to the variability of many other cinchona extracts and tinctures on the market. Squibb saw Stearns's new preparation as a direct offense against the integrity of the association and called for his expulsion.[60] A few members called for just the censure of Stearns. Others suggested that Stearns simply withdraw the product from the market, but he refused.[61]

In the midst of the lively debate, a number of members pressured Procter for his opinion. Although reluctant to speak on the question of expulsion, Procter commented on Stearns's defense and then said:

The light in which I look at this thing is that an Association of this kind, to be of use in the future, must have virtue in our actions. If there is not virtue in us we might as well disband. I look upon this action of Mr. Stearns as a stab at that virtue; as a deep wound to it, and that the action we now take must be in reference to the cure of that wound; if we can cure it without cutting off our fellow member we have had so much occasion to love and respect, let us do it, but if we cannot cure it without using the knife, let us do that.[62]

The assembled body "chose the knife," and discharged Stearns by a vote of sixty-two to twenty-three.[63] Stearns responded: "[I have] been hastily expelled from this Association . . . judge the future by what I do." Ejected from the association for misbranding, Stearns later initiated the first full line of pharmaceuticals manufactured in the United States that were precisely labeled showing all ingredients. Through this innovation, Stearns recovered his good name and standing within the association.

For many years Procter, like most of his contemporaries, had argued that cases like that of Stearns were best handled within the profession itself. Although Procter tolerated taxation of pharmacy practice by state and local governments, he had opposed direct regulation of pharmacy practice throughout most of his career. The association's members periodically addressed the issue of pharmacy law, changing their posture dramatically from 1852 to 1872. In the 1850s much of the impetus for pharmacy practice laws came from physicians concerned with the quality of medicines available from many drug sellers. The issue of poison dispensing produced the most agitation, and when physicians called for laws requiring prescriptions for all poisonous drugs, some pharmacists reacted strongly.[64]

Other members of the association, such as Edward Parrish, took a laissez-faire attitude: "I think that all laws and legislation, whether among ourselves or elsewhere, that tend to confine

Procter's successor as editor of the *American Journal of Pharmacy,* John M. Maisch (1831–93) was born in Germany and came to the United States in 1848. Like so many other pharmacists with continental roots, Maisch emulated the scientific achievement and social standing of European pharmacists. He practiced and taught pharmacy in both New York and Philadelphia. (From Kremers Reference Files, University of Wisconsin-Madison School of Pharmacy, Madison, Wisc.)

our profession, and keep it back, should be carefully guarded against."[65] Procter, long a supporter of the drug importation law, took the more moderate position that pharmaceutical societies should introduce their own limited laws, primarily controlling poison sales and calling for registration of pharmacies and pharmacists.[66]

Fear that pharmacy would be regulated from outside the calling caused the association, in 1868, to appoint a special committee to draft a model law for consideration by state legislatures. Surprisingly, the association's permanent secretary, John Maisch, was originally left off the roster of this committee and only added at the last minute. He later composed the model act that eventually became the basis for laws passed in a number of states.[67]

At the next meeting, Maisch read his "Draft of a Proposed Law to Regulate the Practice of Pharmacy" to the assembled body. It touched off a heated, disjointed debate. Procter got the floor and spoke against a motion calling for each section of the new law to be analyzed. Why waste all the time of the meeting, he reasoned, when the principles of the law matter so much more than the details. Specifics would have to be changed anyway to make up for the different practice conditions in eastern and western states. Without a second, however, his motion died. The discussion began, with Squibb dissecting the preamble. Procter regained the floor and, through some parliamentary wrangling, got the debate shifted from details to principles.[68]

With that settled, Procter reduced the controversy to one major question: Should registration of practitioners by the state be the form of regulation endorsed by the association to eliminate the unqualified? After considerable discussion and numerous amendments by Squibb, the assembled members finally decided to package the model law and association resolutions into a pamphlet and send it off to governors and state legislatures.[69] Procter's ability to focus the association's discussion again proved valuable. On his return to Philadelphia, he furthered the cause of pharmacy law by opening up the pages of the *American Journal of Pharmacy* to discussions of Maisch's model law.[70]

By 1871 Procter's attitudes toward pharmacy law had come full circle. While other leaders, like Parrish, Squibb, and Colcord, maintained that the free market adequately regulated the professions, Procter supported modest legal reform. He never clearly explained his change of mind in print, but his firsthand glimpse of strictly regulated continental pharmacy in 1867 probably played a part. In addition, in the late 1860s and early 1870s Procter's relationship with John Maisch grew from occasional correspondence to daily contact.[71]

Procter's attitudes concerning pharmacy law changed over the years, but his strong opinions on the national character of the association never wavered. The first few annual meetings of the association were eastern affairs by necessity: pharmacy was better developed on the Atlantic Coast, and transportation to the West, while improving every year, frightened off many an easterner. In the mid-1850s Procter tried to stimulate interest in western meetings as a way to make the association truly national. He described potential trips to Cincinnati, Detroit, or Chicago in romantic terms. In Procter's editorials, pharmacists read of the beauties of Niagara Falls, not the innumerable railroad transfers and roads that alternated between choking dust and impassable mud.[72]

In the late 1850s, like other authors with a national audience, Procter preached the theme of unity. At association meetings, he pushed hard for sites that would allow southern members to attend, such as those close to north-south rail or shipping lines. In 1858, for instance, he lobbied for Washington, D.C., as a site and succeeded.[73]

The Civil War brought special problems. The St. Louis meeting had to be cancelled and few members attended the 1862 Philadelphia get-together. Procter hoped to solidify the loyalties of border state members in Maryland and Delaware by holding the 1863 meeting in Baltimore.[74] With the conclusion of the war the association, spearheaded by Procter, welcomed southern members back to the fold.[75] To strengthen the new unity of the association, Procter suggested, in 1869, that the next meeting be held in Richmond, Virginia. The site selection committee had offered Saratoga Springs as a possible site in 1870. Squibb agreed with Procter that this sounded like a vacation, not a working meeting. As a compromise, the association chose Baltimore as the 1870 site.[76]

The Baltimore meeting of 1870 helped to broaden the association into a national society. Much to Procter's pleasure, many southern members, new and old, attended the meeting. In addition, the association elected a Virginian as president and a Georgian as vice-president.[77]

Now Procter wanted a meeting held in one of the former Confederate states. It took two years of arm twisting and pages of editorials, but in 1873 Procter attended in Richmond what was to be his last meeting and was treated royally by grateful pharmacists of that area. His dream of a unified, national pharmaceutical association, first conceived over twenty years before, had been realized.[78]

Returning to Philadelphia from Richmond, Procter delivered his delegate report on the meeting to the College of Pharmacy. With apparent pleasure he reported that many members of the association made the trip south, several with interesting papers in hand. In the early years of the association, Procter would have ended his report there, but in 1873 he continued: "A large portion of the visiting members came in the steamer from Norfolk, and were met a few miles below the city by the Committee of Reception from the druggists of Richmond on a barge, and conveyed to the landing, and thence to their hotel in stages. The welcome extended was most warm and friendly, and through the entire continuance of the session in various ways."[79]

One must remember that these annual meetings were independent affairs, with their own style, pace, and rhythm. A small number of ambitious and sober men attended the first few. As the association grew, more ordinary pharmacists attended and joined in the sessions. They saw the meetings as social as well as scientific and professional events. Host pharmaceutical societies, in the late 1850s, began elaborate banquets and excursions. At first Procter and other founders of the association reacted against this "extravagance." In their view, any excursions should be professional in nature, such as visiting a drug manufacturer.[80]

By the tenth meeting in Philadelphia (1862), however, Procter's attitude toward excursions had changed. At the conclusion of the final session, Edward Parrish invited both friends and strangers to go up the Delaware River to his home in Riverton. In an editorial, Procter described glowingly the sail and subsequent party.[81] His description of the 1866 gathering in Detroit contained this touching passage: "Late in the evening, after the adjournment, a number of members were congregated in the rotunda of the Hotel; they were about to depart to their distant homes . . . some by the night trains, others in the morning. Suddenly, as by one accord, the idea of *"Auld lang syne"* seized on the group, and, joining hands in a circle of twenty feet diameter, the heart-stirring tones of the old song were brought out with an emphasis and feeling that will long be remembered."[82]

The lack of serious internal disagreement within the association partially explains this high level of camaraderie. The association members—as evidenced by the words of the leadership and the stenographic records of the sessions over the years—united behind an overriding purpose: to raise the practice of pharmacy in the eyes of physicians and the public above the level of an ordinary trade.[83]

Disagreements did sometimes arise, with several issues producing considerable controversy in the 1850s, 1860s, and 1870s. Many annual meetings were recorded by a stenographer, preserving some of the flavor of those debates, including Procter's own role. Procter, as discussed above, would usually let the hotheads speak their minds and then inject his own analysis of the problem at hand. For instance, at the 1868 meeting, the issue of taxes and tariff took up considerable time at one session. Alonzo Robbins had spoken for several minutes on the evils of the tariff, when he yielded the floor to Procter for a question:

PROCTER: Is not the tariff to pay the interest on the government debt?
ROBBINS: Yes, sir.
PROCTER: Then let us pay it.

This abrupt response caused Robbins to abandon his line of argument on the tariff and shift to other concerns.[84]

Parrish and Procter, two old friends of like minds, rarely disagreed at association meetings, but the possibility of hosting the

International Pharmaceutical Congress in 1876 divided them. Procter, who had attended the congress held in Paris in 1867, saw it as a waste of time. In his view, delegates just sat around talking local pharmacy politics. The matter came to a vote and only one other member sided with Procter—a good example among several that Procter did not always get his way at association meetings.[85]

In other disagreements between these two association leaders, Procter usually won out. Parrish, in contrast to the reflective Procter, had a quick wit and used it often at association sessions. He was an impatient man, pushing the association ahead at the same time that the more conservative Procter advocated more gradual growth. Recognized by his peers as an excellent speaker, Parrish tended toward verbosity. His presidential address may be the longest on record, three times longer than Procter's. A man whose pharmaceutical interests rested in the practical rather than the theoretical, Parrish oddly enough injected more philosophical allusions into his rhetoric than any other member of his time. But a sharp tongue apparently alienated other members of the association and decreased his influence on its policies.[86]

Another man with firm beliefs who rarely disagreed with Procter was Edward R. Squibb of New York. Deeply respected by members of the association for his ethical manufacturing standards and scientific abilities, Squibb had a temperament between Procter's and Parrish's. His attention to detail sometimes bogged down association discussions, yet it also tended to focus debate. Both a physician and a pharmaceutical manufacturer, Squibb often changed hats in the midst of a controversy, when one or the other profession needed a defender. Just as earlier physicians like George B. Wood and Franklin Bache helped pharmacists get into the pharmacopoeial revision process, Squibb's moderating influence played a large part in the respectful notice the association received from other related organizations.[87]

In Procter, Squibb found an enduring friend and colleague. After a disfiguring accident at his Brooklyn plant, Squibb did little socializing. With Procter, however, he felt at ease. They shared several association committee assignments over the years, producing key reports dealing with drug quality and pharmacopoeial standards. In association matters, Squibb, ever the practical capitalist, spoke for the economic well-being of pharmacists; Procter, the idealist, championed professionalism. Together they formed a team that set much of the course for the young American Pharmaceutical Association.[88]

Where did Procter fit into the leadership scheme? He was most prominent during the birth and infancy of the association, molding its early character. After the first ten years or so he stepped back a bit and let others, like Squibb and Maisch, provide new direction for the organization. Procter became more of an overseer, a judge of new proposals. He acted as a sort of guardian of the association's mission. His involvement in the founding earned him respect from members in later years and added to his influence; his apparent lack of personal ambition strengthened the moral authority of his positions.

In summary, Procter viewed organizations as vehicles for the self-improvement of individual members. By raising the level of the practice of each pharmaceutist, the public and the profession itself would benefit. His approach contrasted with that generally held by other leaders of the young profession who contended that pharmaceutical organizations aided the individual practitioner best by raising the status of the calling as a whole.[89] "We do not mean," said Procter, "that an apothecary, by being a member, will add a certain number dollars to his income,—possibly it may not increase it a dime. But we do mean that it will open a field wherein he can glean much that may be improving to himself, and at the same time, without loss, enable him to extend a useful influence to others less enlightened and successful than himself."[90]

8 Concluding Remarks

There is a force in the Anglo-Saxon character, which, if once brought to bear on any object, will gain its end, or bring about material changes in the condition of the things opposed to it; and the results of its energetic manifestations in favor of pharmaceutical reform have been greatly proportioned to the unanimity of sentiment, and combination of *will*.[1]

With these words Procter, an avid admirer of British pharmacy and culture, provided an excellent description of his own unique character. In several of the challenges that Procter took on he did not gain his end, but did bring material changes. As a small businessman he never made a fortune, but did show that one could live by practicing prescription-based pharmacy. As a pharmaceutical scientist he made no earth-shattering discoveries, but did stimulate the scientific endeavors of various students and colleagues and provided an open forum for investigations in the *American Journal of Pharmacy*. As a participant in pharmacopoeial reform he was frustrated by the exigencies of war and the subsequent growth of pharmaceutical industry, but did help to set the precedent of pharmaceutical participation in the setting of drug standards that eventually led to the adoption of the *United States Pharmacopoeia* and the *National Formulary* as legal compendia. As a pharmaceutical editor he struggled to keep his periodical viable in difficult times, but was able to hand over to his successor a healthy *American Journal of Pharmacy* that remained a recognized leader in its field for the next century. As a pharmaceutical educator Procter reacted against some of the changes brought on by the industrial revolution and laboratory-based science, but did demonstrate the value of formal instruction in practical pharmacy, which fed into the eventual development of an academic discipline of pharmaceutical chemistry. As a founder of the American Pharmaceutical Association Procter did not succeed in molding that organization in the image of the Philadelphia College of Pharmacy, but did help put the APhA on a course that made it a major force in American pharmacy.

One could argue that Procter did not do anything spectacularly different from some of his colleagues involved in pharmaceutical activities during the mid-nineteenth century. Before him men like Daniel B. Smith and Elias Durand carried the banner of pharmaceutical science and art. Procter's contemporaries Edward Parrish and Edward Squibb wrote on ways to improve pharmaceutical practice and drug preparations. Yet Procter stands above any other figure in American pharmacy because of the size and scope of his contribution to his calling. He and his famed protégé, Joseph Remington, were among the last major figures in American pharmacy to be complete generalists, that is, run a drugstore, teach at a college, do pharmaceutical research, write, and perform a wide variety of organizational duties. After Procter's students had been supplanted by the next generation of pharmacists, American pharmacy had progressed to the point where specialization was necessary for an individual's success.

For his efforts, William Procter has been canonized in numerous speeches, awards, paintings, and sculpture. A life-sized bronze statue of a seated Procter, bearded and robed, dominates the rotunda of the headquarters of the American Pharmaceutical Association. The statue was sculpted by William F. Simpson and unveiled at the American Institute of Pharmacy on May 3, 1941. Plans for the statue had begun in 1899.[2] The pedestal of the statue reads "Father of American Pharmacy," a durable epithet bestowed upon Procter by pharmacists soon after his death in 1874.[3] Even

Seated in robed splendor, a life-sized bronze statue of William Procter, Jr., holds court in the rotunda of the headquarters of the American Pharmaceutical Association in Washington, D.C. (From *Journal of the American Pharmaceutical Association* 2 [1941]: 168.)

though modern historiography shuns such titles, their use is an indicator of the heroes of the past. For American pharmacy, Procter has served as a symbol of identity and pride.

For pharmacists of the late nineteenth century, concerned over the apparent threat that the rise in factory-made medicines posed for their prestige, Procter represented an ideal, albeit outmoded, that placed special virtue in medicaments made by the pharmacist with his own hands. For example, Joseph Remington and other pharmaceutical leaders asked physicians to compose individualized prescriptions for each patient from basic ingredients stocked in pharmacies. Even though the pharmacopoeial ingredients would be purchased through wholesalers or from large-scale manufacturers, the pharmacist's skill would be needed for these "tailor-made" prescriptions.

For the pharmacists of the early twentieth century who had received formal instruction in basic science, Procter exemplified the "chemist on the corner" image that many wanted American pharmacy to project. In addition, Procter's dedication to pharmaceutical organizations made him an attractive hero for these groups to rally around. During the 1920s, for example, solicitations for contributions to the new headquarters of the American Pharmaceutical Association included a proposed memorial to Procter.[4] In the 1930s and 1940s the colorful president of the Philadelphia College of Pharmacy, Ivor Griffith, revived interest in Procter through several addresses, articles, and short sketches. For him Procter served as an excellent symbol of the notable achievements of his institution. Since the 1960s, however, pharmacists' memory of Procter has faded, perhaps because of the near disappearance of compounding in pharmacy, the decline in the profession's identification of the independent store owner as its model practitioner, and the reduced role of heroes in the American ethos.

In previous chapters I have not said much about the impact that Procter and his work had on the public welfare, primarily because it was hard to gauge in the context of the individual subject areas discussed. Moreover, because he did not develop or discover an important drug or design a new dosage form, his broader influence on the public is harder to visualize. As a practitioner Procter ran a small business whose direct influence did not extend beyond his neighborhood. But he practiced exceptional pharmacy that foreshadowed some future developments. He did not practice medicine over the counter or sell poisons without prescriptions, which were then common practices. He kept on hand freshly prepared antidotes for poisons, which saved several lives. As a pharmaceutical scientist and reviser of the *Pharmacopoeia* Procter contributed to the well-being of society by improving extraction techniques and other preparation methods, which increased the predictability of the effects of a medicament from one batch to another.

As a model for the next generation of pharmacists, Procter perhaps contributed most to the public welfare. American pharmacists during the late nineteenth century gradually abandoned the selling of nostrums in their shops, substituting preparations of known formulation. Procter's encouragement of this approach probably played a part in the eventual publication of the *National Formulary* (1888), a compendium that attained official status in 1906.[5] Moreover, he championed the idea of a responsible pharmacist who had the education and self-confidence to act as a safety check between the writer and the consumer of prescriptions. Procter deserves much of the credit (or blame) for the pharmacist being cast as a sort of critical underling to the physician: entrusted with questioning but never overriding his superior's orders.

Procter liked to tell the story of how he and a fellow apprentice battled on his first day at Zollickoffer's shop with large enema syringes filled with water drawn from the gutter. Tiring of this escapade young William returned to the apothecary shop and picked up a copy of Henry's *Chemistry,* commenting, "This is just what I like."[6] The amount of truth in this anecdote is unknown; yet it tells us something about Procter's self-image. He enjoyed looking back to his years at Zollickoffer's shop, when he spent hours studying the wonders of the natural world and the laboratory bench, sharing his enthusiasm for the future with young colleagues like Augustine Duhamel. Procter saw science as a path to higher truth through individual effort, much as his Quaker upbringing had suggested a personal approach to God. Science seemed to be a key to the progress of American society and a way to elevate his mundane trade to the level of a respected calling.

When Procter's close friend Duhamel died in his prime, Procter composed a verbal salute that could well serve as his own epitaph:

"The character of his mind, though not marked with striking originality, was particularly well suited for his profession; and in viewing subjects he was influenced more by the practical and useful, than by any abstract scientific interest they might possess. . . . His mild and courteous demeanour marked him as a gentleman, and in his intercourse with society the impressions left on the minds of his associates, were altogether in favour of the goodness of his heart. May we who are left emulate his virtues."[7]

Appendixes

Appendix A:
Procter's Paris Speech

[In 1867 William Procter attended the International Pharmaceutical Congress held in Paris as a delegate from the American Pharmaceutical Association. After listening to some continental pharmacists degrade their American counterparts, Procter composed and read the following speech. This translation by George Urdang has not been published previously.[1]]

The discussion which has taken place at our first session, and of which I am sorry to say that I did conceive little—incapable as I was to grasp its definite meaning—dealt with a subject little known in America, namely governmental interference with the practice of pharmacy, be it by the requirement of a diploma or by the limitation of the number of practitioners based on the population, or by any other means of control over the profession of pharmacy.

I have become aware that, on the one hand, the double benefit which may result for the public interest as well as the profession of pharmacy (hence the convenience of such an interference) may be discussed, and that, on the other hand, many doubts manifest themselves concerning the feasibility of a system excluding competition as an element of progress.

The experience of a country in which pharmacy is as little

shackled by the law as it is sustained by the government, of a country where the empirics [quacks] are held in the same public esteem as the pharmaceutists, has in the opinion of some persons offered a source from which to derive arguments for or against the strict rules in use in several European countries. If my information is correct, the method of inquiry has had the effect to give a wrong idea of the real state of pharmacy in the United States. It is for this reason that I think it my duty to present a short statement in order to make the real position of America with regard to this question evident. In fact, whatever opinion one may have about the situation of pharmacy in this country, this situation is the natural result of the circumstances from which it originated and developed.

Only a few years ago pharmacy as a profession or as a distinct trade of its own did not exist, its functions were taken care of by the physician and his aide; consequently the prescriptions of the physician were filled by himself or his apprentices, not being submitted to the control of a responsible pharmaceutist.

Today, however, there exists a new order of things: the physician has gradually and, by now, almost entirely ceased to play the part of the pharmacist.

The change has taken place in the Atlantic states and will continue until the wildernesses of the West have been filled like these states are today, in hundreds of cities, of which several count 200,000 inhabitants. In the beginning the pharmaceutists who enjoyed the confidence of the physicians were only a small number, and their establishments were reputed because of their skillful manipulation and the good quality of the remedies which they delivered. When they were in need of a greater number of aides, young people were in this way initiated into the practice of the store, which thus became a sort of first school of pharmacy, and it was there that some of the best pharmaceutists of our days obtained their professional education.

About 1821 some of these pharmacists, in the interest of numerous young men placed under their direction and with very limited theoretical background, founded the Philadelphia College of Pharmacy, today an institution recognized by a charter, and organized at the same time courses in chemistry and materia medica. Six years later, the first graduates obtained the diploma which the College by its charter was entitled to confer.

In the opinion that the store and its laboratory were, during the duration of an apprenticeship, the veritable school for instruction in practical pharmacy, these first promoters of our art did not dream of the establishment of a school for practical pharmacy, and it was not until 1846 that, in Philadelphia, instruction of this kind was created in the college of this city in which, under the eyes of the students, experiments were pursued as much as possible.

This instruction did not include and does still not include personal experimentation by the student in the laboratory under the supervision of the professor. This latter and more efficacious method of familiarizing the student with practical pharmacy has been adopted in the United States only in restricted limits, primarily because of the difficulties offered by the lack of time on the part of the students.

This statement applies to all colleges of pharmacy and, although the necessity of a change may be admitted as extraordinarily desirable, the time has still not come to demand, with sufficient authority, the creation of laboratories for practical work in the colleges. It results from this state of affairs that, although the word *student* applies to all, those students distinguish themselves through a number of divers qualifications who have become familiar with important laboratory details with the help of their professors or by their individual industry.

Some only learn the [drug]store service. If one takes the term *store service* in its real sense, their number is limited. But many have worked in a way that they may serve as an example to the students of the schools so well organized in the old European countries.

There exists an ardent desire to learn pharmacy, and it is increasing. Each year, young people come from the interior to the schools of Philadelphia and New York to pursue their studies; and the last winter class of the Philadelphia College of Pharmacy had 160 students.

The young men who have been educated in those schools and have established themselves all over the country have carried with them the germs of a better condition of our art and have exercised a fortunate influence on pharmacy through their example and the emulation which it has excited in others.

Public opinion in America is a mighty agent of reforms and it has been the main source of progress in pharmacy. The graduates of the colleges of pharmacy enjoy, by virtue of their diplomas, a reputation which gradually makes itself felt and promises a better future.

Another source of improvement has been the immigration of pharmaceutists who have come to us from Europe and especially from Germany. Some among them have exerted a remarkable influence and have aided in the elevation of the standing of pharmacy. Frequently, however, they have addressed themselves more specifically to their compatriots residing in America, supporting them with medicaments prepared according to European custom.

One can ask oneself, what is the condition in pharmacy brought about by this free competition of the profession and this absence of any law limiting the number of pharmacists?

There is in fact not the slightest obstacle toward a multiplication of drugstores save that of a lack of success. The number of pharmaceutists existing in a given place is of little importance to another druggist who thinks himself able to satisfy the public better. He will try, and if he succeeds, some of those who had been established before him may be forced out of business, at least if the simultaneous growth of population caused by immigration does not make a greater number of shops necessary. In this way, competition acts as a means of improvement. In the eyes of those accustomed to the Old World system, this practice may seem rather cruel toward those already in business. It is, however, the only one permitted by present-day American public opinion with regard to the creation of special privileges, and in a country where so many opportunities are open to everyone to make a living, there are no difficulties in changing career.

The Pennsylvania legislature will accord a charter to any group of citizens making known its intention to teach medicine, pharmacy, or homeopathy. It will, however, not vote a law forbidding other citizens to practice medicine or pharmacy, and it will not prevent any person, doctor, pharmaceutist, or empiric to adopt whatever title seems good to him. Everybody is subject to the law for bad practice, and the law applies quite as well to the negligence of the surgeon as to the ignorance of the charlatan.

In thus giving an idea of the present state of pharmacy in the United States, I have thoroughly reserved my personal opinion. If asked for that, I am in favor of a system which allows the practice of pharmacy only to graduates of regular pharmacy schools and leaves it definitely to the power of competition and to the peculiarities of the profession to regulate the number of those practicing pharmacy.

Since we were asked to vote for or against an international pharmacopoeia, we are compelled to answer in the negative because of our belief that the ideas of English and American pharmacy are, according to many reports, so different from the continental European, that they resist a unification on this subject except as far as they touch inorganic drugs.

In England as well as in America the physicians have had, and still have, the main influence on the editing and preparing of the pharmacopoeias. In the United States, until 1850, the work of pharmacopoeial revision was entirely entrusted to the physicians. In that year and in 1860 the national conventions meeting in Washington, [D.C.,] were composed of delegates of all medical and pharmaceutical groups forming regular corporations, and having been a member of these conventions at both occasions, I can confirm on the basis of this personal knowledge that each new proposition having as its aim the adoption of a pharmacopoeia conforming with the [French] codex or the Prussian pharmacopoeia, would meet an unsurmountable barrier.

As far as the weights and measures are concerned, it is different. There has manifested itself, among the American men of science, a growing interest in favor of the metric system and in Washington, [D.C.,] last winter, after the examination of the question by a committee of the National Academy of Science, the Congress of the United States has voted a law rendering the use of the metric system legal, but not obliging anybody to adopt it and to cease to supply the avoirdupois and the measures for wine.

The American Pharmaceutical Association has, last year, expressed a favorable opinion toward the decimal weight system which eventually may be adopted. I say that this is a result likely to be realized only in the future, since I know the obstinacy with which people stick to old customs. Personally, I am aware that the "avoir du poids" offers some conveniences to the ordinary commerce that it will be very difficult, in practice, to have its use stopped in England and in America.

At this occasion, I have to report that Mr. Alfred B. Taylor, pharmaceutist in Philadelphia, has sent me a brochure about the octaval system, the adoption of which he suggests, together with a steel measure, representing the middle of a length-unit adopted in his system. Last March when I passed through Paris, I left everything in the trustworthy hands of the venerable president [Nicholas] Guibourt in order to be shown to the Congress and afterwards to be presented to the Society of Pharmacy. I presume that the society is today in possession of these two objects.

In asking you to accept my apologies for having made the Congress lose so much time, I conclude respectfully.

Appendix B:
Examples of Procter's
Writing Style

[Procter was well known for turning an interesting phrase. What follows are a few examples of his prose, with introductions when necessary.]

"Professorial Metamorphosis"

The curious changes which the insect undergoes in obedience to the laws of its being, are occasionally to be observed among other classes of animals. For instance, a tadpole may turn to a frog, and even a poor half-starved doctor, or apothecary, after a combat with the pressure of circumstances, has been known to gradually assume the *status empiricalis,* but it has been only quite lately discovered that a regular medical professor, with all the collateral titular insignia of professional importance, is capable of sudden transformation into a full grown quack, amply provided with the material for generating and the organism for disseminating those gaseo-literary exhalations so perculiar to the latter individual. Such changes among the lower animals are generally progressive; the unsightly grub, and the uncouth tadpole, becoming the beautiful butterfly, and the active symmetrical frog; but in the case of the professor, the movement is decidedly retrogressive, probably from the fact that the change is abnormal.

The true cause of this remarkable phenomenon is not generally known. Some have supposed it a case, *sui generis,* never before described; others regard it as arising from a diseased condition of the lateral portions of the cerebrum; whilst a third class of investigators deny that any metamorphosis occurred, and consider that there is nothing remarkable in the case, except that obliquity of vision in public perception, which so long caused a quack to be mistaken for a true professional man.[1]

"Gelseminum *versus* Jasminum"

[A physician named Brooks made the contention that jasmine was known to the ancient Romans, to which Procter responded:]

The learning of Dr. Brooks has for this once, at least, proven false to him. The learned old Greek [Dioscorides] wrote about many things, some of which it is difficult now to recognize, but he never wrote about or heard of the *Gelseminum sepervirens* of our Southern States, which no doubt flowered and evolved its fragrance in the days of Nero, as it does now; but Nero's galleys never crossed the Atlantic to bring that wise physician this valuable American plant.[2]

Book Reviews

[Procter reviewed scores of books. Here are three short excerpts. The first deals with a pamphlet describing a new gold-mining firm in Colorado, the second with Kost's *Elements of [Eclectic] Materia Medica,* and the last with the *Favorite Prescriptions of Living Practitioners.*]

The above pamphlet of sixty-seven pages is an elegant specimen of Boston typography. The statements and illustrations of the pamphlet are plausible, and if the Company's metallic lodes prove as substantial as their paper and printing they will reap a rich return from their mining operations.[3]

The author [of *Elements of Materia Medica*] has been careful to keep beyond the reach of the law of copyright in making copious quotations from Royle and Pereira, which are interspersed through his pages, like patches of silk on garments of cotton.[4]

The book [*Favorite Prescriptions*] is neatly printed on good paper, and will doubtless be valued by that numerous class of medical practitioners who prefer the prescriptions of others to the trouble of adapting medicine themselves.[5]

Appendix C: Example of Procter's Work in Drug Assay

"Note on Adulterated Powdered Socotrine Aloes"

Having been recently called upon to examine two specimens of aloes in reference to their purity, it has been thought that a few remarks relative to the result will not be without use.

Specimen A presents the form of masses of a conglomerated powder, of coarse texture, with numerous minute woody-fibres disseminated through it, and visible to the naked eye. When breathed upon, it exhales the odor of true socotrine aloes; it gives a yellowish-brown slightly greenish colored powder, very gritty under the spatula, and evidently consists of that variety of aloes, admixed with other matter, partly ligneous, and partly inorganic.

Specimen B is a mass and fragments, rendered friable by a thorough dessication, effected with a view to powdering, and readily brakes [sic] up when handled. Its color in mass is reddish brown, modified by fissures on its surface; possesses the odor of true socotrine aloes, and affords a deep orange yellow powder when recently triturated.

The specimen A is part of a lot of about 1000 pounds sent to a Western druggist from New York, and refused on the ground of its not being of good quality. The druggist in New York asserts that the aloes was pure when sent to be powdered, as a retained sample, which was submitted to an analytical chemist in that city, was pronounced to be of good quality.

The specimen B was a sample of a lot of aloes powdered and sent to the same Western druggist by another party, and which proved satisfactory.

In the following statement relative to the treatment of 100 grains of each specimen, it must be understood that each result was obtained from a separate quantity, and not from the same portion.

Specimen A

100 grs. contained 42 grains, insoluble in cold water.

100   "   "   33 grs., insoluble in boiling water.

100   "   "   10 grs. apotheme or altered aloein, deposited from the solution in boiling water by cooling.

100   " yielded 13 grs. of incombustible ash when incinerated in a platina crucible.

33 grs. of residue insoluble in boiling water yielded 12 grs. of incombustible ash by incineration.

100 grains yielded 37 grs. insoluble in alcohol 35° Baume.

Specimen B

100 grs. contained 33 grs., insoluble in cold water.

100 grs.   "   20 grs., insoluble in boiling water.

100 grs.   "   12 grs. apotheme or altered aloein.

100 grs. yielded 6/10ths gr. incombustible ash by incineration.

20 grs. of residue insoluble in boiling water, yielded ½ gr. of incombustible ash by incineration.

100 grs. yielded 2½ grs. of matter insoluble in alcohol 35° B.

The residue of A, insoluble in cold water when examined with a lens, exhibited shreds of wood-fibre, shining gritty particles, and dried pulpy matter in fragments.

The residue of B, insoluble in cold water, had a resinous kino-like look, and consisted of 20 parts of resin and 13 of apotheme.

The large percentage of inorganic matter in A, does not belong to fair commercial aloes, and it is difficult to account for such an adulteration in view of the less ready detection of cheap aloes. The uniform manner in which the woody-fibre is distributed through the specimen A, leads to the supposition that some vegetable powder was introduced with a view to prevent the conglomeration of the aloes; and the large residue insoluble in alcohol is additional evidence of the impurity of the specimen A. Whether the original aloes was very inferior, and contained both woody and earthy matter, or whether designedly adulterated, it is, of course, impossible to determine by analysis.[1]

Appendix D: Example of Procter's Pharmaceutical Art

"On Fluid Extract of Rhubarb and Senna"

Notwithstanding that two preparations of Rhubarb and Senna are already known, it is believed that the new one, now proposed, possesses sufficient claims to gain for it the favorable opinion of physicians and patients in many cases where a cathartic is needed, simply as such, or in connection with other medicines. It is well known, that senna has little, if any, tonic influence on the alimentary surfaces; that an overdose has a depleting effect, often inconvenient, and that griping is a frequent attendant on its exhibition. On the other hand, it is equally understood, that rhubarb is remarkable for being a sort of therapeutical paradox, in so far as it possesses both a purgative and an astringent property, the latter coming into play *after* the former has manifested itself, and thus repairing, as it were, its effects. It is also well known, that this astringent or tonic action is so strongly marked, that it is necessary in most cases to combine it with some other cathartic to overcome or modify this peculiarity when a simple cathartic is needed. By the union of these two drugs in the concentrated form presented by a fluid extract, and in a due proportion, a resulting cathartic action is obtained which is safe, unattended by unpleasant symptoms, and not followed by constipation when the dose has been properly graduated. It has been ascertained that the association of

195

alkalies and alkaline salts with rhubarb and senna, has a tendency to prevent their unpleasant griping effects, and in the case of senna, to increase its activity. The introduction of the bicarbonate of potassa is with this view, and the aromatics from their carminative properties also aid. The following is the formula:—

Take of Senna, in coarse powder,   twelve ounces,
       Rhubarb, in coarse powder,   four ounces,
       Bicarbonate of potassa,   half an ounce,
       Sugar,   eight ounces,
       Tincture of ginger,   a fluid ounce,
       Oil of cloves,   eight minims,
       Oil of aniseed,   sixteen minims,
       Water and alcohol, of each a sufficient quantity.

Mix the senna and rhubarb (by grinding them together in a convenient way,) pour upon them two pints of diluted alcohol (U.S.P.), allow them to macerate twenty-four hours, and introduce the mixture into a percolator furnished below with a stopcock or cork to regulate the flow. A mixture of one part of alcohol and three of water should now be poured on above, so as to keep a constant but slow displacement of the absorbed menstruum, until one gallon of tincture has passed. Evaporate this in a water bath to eleven fluid ounces, dissolve in it the sugar and bicarbonate, and after straining, add the tincture of ginger, holding the oils in solution and mix. When done the whole should measure a pint.

Remarks. If the percolation has been properly conducted, the ingredients will have been sufficiently exhausted when six pints of fluid have passed. As by far the larger portion of the soluble matter passes in the first two pints, it is well to set these aside and evaporate them separately to six fluid ounces, subsequently adding it to the other liquid when it has been reduced to five fluid ounces. As the cathartic principles of senna and rhubarb are very susceptible to injury from heat, especially in contact with the air, the propriety of using the best available means for conducting the evaporation need not be urged. When the evaporation is conducted in open vessels, some advantage is gained by adding the sugar to the tincture and continuing the process until it measures fifteen fluid ounces. The sugar protects the extractive matter from oxidation and more completely suspends or dissolves the resinous part of the rhubarb contained in the tincture. The bicarbonate should not be added to the extract while it is above 140° Fahr. and should be reduced to powder previously.

It may be objected to this formula that we already have fluid extracts of rhubarb and of senna of the same ratio of strength, and that when physicians need such an association, they can mix them. In answer, it may be stated that the cases where a simple cathartic is needed, are so numerous that this preparation will be found useful to the physician, and a good medicine for travellers and others who resort to this kind of purgative habitually.[1]

Notes

Abbreviations

AJP   American Journal of Pharmacy
Am J Pharm Ed   American Journal of Pharmaceutical Education
Drugg Circ   American Druggists' Circular and Chemical Gazette
Proc APhA   Proceedings of the American Pharmaceutical Association

Preface

1. William R. Fisher, "A Brief Sketch of the Progress and Present State of Pharmacy in the United States of America," *AJP* 9 (1837): 271–79.
2. [Charles Bullock], "Memoir of Prof. William Procter, Jr.," *AJP* 46 (1874): 512–33; Joseph P. Remington, "A Memorial of William Procter, Jr.," *Proc APhA* 48 (1900): 22–28; Remington, "Procter Reminiscences," *Proc APhA* 54 (1906): 549–61.
3. According to Nathan Singer of Sherman Oaks, California, Ed Zaslow, former drug inspector for the state of New Jersey, claimed to have thrown out Procter's pharmacy records from the attic of an old Philadelphia drug shop around 1920 (personal communication, July 1982). In addition, Dean Emeritus Linwood Tice disposed of Procter's (and Joseph Remington's) lecture and research notebooks when he began teaching at the Philadelphia College of Pharmacy (personal communication, September 1983).

Introduction

1. The following is based largely on my article, "Professionalism and the Nineteenth-Century American Pharmacist," *Pharmacy in History* 28 (1986): 115–24, in which I discuss the issues in more detail.
2. One classic example of this genre is *The Professions,* by A. M. Carr-Saunders and P. A. Wilson (Oxford: Clarendon Press, 1933).

3. Eliot Freidson, "The Theory of Professions: State of the Art" in *The Sociology of the Professions,* ed. R. Dingwall and P. Lewis (New York: St. Martin's Press, 1983), 32–33.

4. Everett C. Hughes, *The Sociological Eye* (Chicago: Aldine-Atherton, 1971), 377.

5. Donald M. Scott, "The Professional that Vanished: Public Lecturing in Mid-Nineteenth-Century America," in *Professions and Professional Ideologies in America,* ed. Gerald L. Geison (Chapel Hill: Univ. of North Carolina Press, 1983), 14.

6. For more on the pitfalls of the attribute model of professionalism, see Julius Roth, "Professionalism: The Sociologist's Decoy," *Sociology of Work and Occupations* 1 (1974): 18; and Robert Dingwall, "Accomplishing Profession," *Sociological Review* 24 (1976): 331–35.

7. Much of what follows comes from research on the practice of pharmacy in Philadelphia, Procter's home. Philadelphia was not typical during this period, but served as a center for medical and pharmaceutical education in the United States and provided models of practice emulated in the West (Procter with Joseph Carson, "Editorial Department," *AJP* 22 [1850]: 189). [All subsequent references to works by Procter will be cited using his last name alone.]

8. Like *medicine, pharmacy* can be a confusing term. It can refer to a profession, an academic discipline, or a shop. Procter defined it operationally as *"the art of preparing Medicines.* It embraces the knowledge and ability that enables an individual to modify those materials of nature, that accident or research have discovered to possess remedial powers, in such a manner as to conduce most effectively to their actions as medicine" (Procter, "Lecture, Introductory to the Course on Practical Pharmacy . . . ," *AJP* 19 [1847]: 245).

9. Guenter B. Risse, "Introduction," in *Medicine Without Doctors* (New York: Science History Publications, 1977), 2–4.

10. Edwin T. Freedley, *Leading Pursuits and Leading Men* (Philadelphia: Edward Young, 1856), 156ff.; Glenn Porter and Harold C. Livesay, *Merchants and Manufacturers* (Baltimore: Johns Hopkins Univ. Press, 1971), 27–36.

11. "James W. Simes, Druggist and Chemist," (Philadelphia) *Public Ledger,* April 6, 1836, p. 1.

12. Gerald Carson, *The Old Country Store* (New York: Oxford Univ. Press, 1954), 237.

13. Ibid., 257.

14. Procter with Joseph Carson, "Editorial Department," *AJP* 20 (1848): 350.

15. [Procter], "Eclectic Pharmacy," *AJP* 26 (1854): 108–9.

16. Alex Berman, "A Striving for Scientific Respectability: Some American Botanics and the Nineteenth-Century Plant Materia Medica," *Bulletin of the History of Medicine* 30 (1956): 17–23.

17. Whitfield J. Bell, *John Morgan: Continental Doctor* (Philadelphia: Univ. of Pennsylvania Press, 1965), 147.

18. For lists of the drugs commonly included in such chests, see the appendix to Edward Parrish, *An Introduction to Practical Pharmacy* (Philadelphia: Blanchard and Lea, 1855). See also W. A. Brewer, "Reminiscences of an old pharmacist #10," *Pharmaceutical Record* 4 (1884): 442.

19. Bell, *John Morgan,* p. 241.
20. Chauncey A. Goodrich, rev., *An American Dictionary of the English Language,* by Noah Webster (Springfield, Mass.: Merriam, 1854), 821.
21. W. A. Brewer, "Reminiscences of an old pharmacist #2," *Pharmaceutical Record* 4 (1884): 232.
22. *A Brief Account of the New-York Hospital* (New York: Isaac Collins and Son, 1804), 3.
23. *An Account of the New-York Hospital* (New York: Collins, 1811), 9–10.
24. *By-Laws and Regulations . . . of the New-York Hospital* (New York: Mahlon Day, 1820), 15.
25. Charles E. Rosenberg, "Social Class and Medical Care in Nineteenth-Century America: The Rise and Fall of the Dispensary," *Journal of the History of Medicine* 29 (1974): 40.
26. Glenn Sonnedecker, rev., *Kremers and Urdang's History of Pharmacy,* 4th ed. (Philadelphia: J. B. Lippincott, 1976), 255.
27. *The Pharmacopoeia of the Massachusetts Medical Society* (Boston: E. and J. Larkin, 1808), v.
28. George Rosen, "Fees and Fee Bills: Some Economic Aspects of Medical Practice in Nineteenth Century America," *Bulletin of the History of Medicine,* supp. 6 (1946): 3–6.
29. "Proceedings of the New York College of Pharmacy," *AJP* 18 (1846): 248ff.
30. W. A. Brewer, "Reminiscences of an old pharmacist #7," *Pharmaceutical Record* 4 (1884): 348.
31. Porter and Livesay, *Merchants and Manufacturers,* 35; Freedley, *Leading Pursuits,* 164–81.
32. Franklin Bache, "Address Delivered before the Graduates of the Philadelphia College of Pharmacy," *AJP* 7 (1835): 93ff.
33. "Introduction," *Journal of the Philadelphia College of Pharmacy* 1 (1825): 1.
34. Fisher, "A Brief Sketch of the Progress and Present State of Pharmacy," 271ff.; Joseph Carson, "Address Delivered to the Graduates of the Philadelphia College of Pharmacy," *AJP* 11 (1839): 100.
35. J[acob] B[igelow], "[Review of] The Pharmacopoeia of the United States of America . . . 1830," *American Journal of Medical Sciences* 8 (1831): 149.
36. Samuel Jackson, "On Belladonna in Pertussis," *American Journal of Medical Sciences* 14 (1834–35): 368.
37. F. B. Kilmer and Charles D. Deshler, "Drug Clerks One Hundred Years Ago," *Journal of the American Pharmaceutical Association* 18 (1929): 714.
38. Anon., "Apothecaries," *Boston Medical and Surgical Journal* 11 (1834–35): 261.
39. Samuel Jackson, "Case of Supposed Poisoning with Arsenic," *American Journal of Medical Sciences* 5 (1829): 237ff.
40. (Philadelphia) *Public Ledger,* June 9, 1836, p. 4.

1. A Glimpse at Procter's Character

1. [Bullock], "Memoir of William Procter," 516ff.
2. Philip S. Benjamin, *The Philadelphia Quakers in the Industrial Age* (Philadelphia: Temple Univ. Press, 1976), 33.

3. [Bullock], "Memoir of William Procter," 516.

4. Ibid., 516–7. Compare Remington, "Procter Reminiscences," 549.

5. [Bullock], "Memoir of William Procter," 516–17. Zollickoffer was a founder of the Philadelphia College of Pharmacy in 1821. Joseph Turnpenny (1812–92) practiced pharmacy in Philadelphia until 1864 and coauthored a few pharmaceutical papers with Procter.

6. W. J. Rorabaugh, *The Craft Apprentice* (New York: Oxford Univ. Press, 1986).

7. Douglas T. Miller, *The Birth of Modern America, 1820–1850* (Indianapolis: Bobbs-Merrill, 1970), 107–9.

8. Merle E. Curti, *The Growth of American Thought,* 3d ed. (New York: Harper and Row, 1964), 341.

9. Although still important, prospective physicians came to rely less on apprenticeship for their education than did pharmacists in the early to mid-nineteenth century. See Martin Kaufman, "American Medical Education," in *The Education of American Physicians,* ed. Ronald L. Numbers (Berkeley: Univ. of California Press, 1980), 15; Paul Starr, *The Social Transformation of American Medicine* (New York: Basic Books, 1982), 42.

10. Glenn A. Sonnedecker, "American Pharmaceutical Education before 1900" (Ph.D. diss., University of Wisconsin, 1952), 275; Bache, "Address," 89.

11. Sonnedecker, *Kremers and Urdang's History of Pharmacy,* 180; Sonnedecker, "Pharmaceutical Education," 28.

12. [Bullock], "Memoir of William Procter," 519.

13. Ibid., 520. Hereafter, the Philadelphia College of Pharmacy will be referred to as "the College" when the meaning is clear. It was a local pharmaceutical society founded in 1821.

14. "Minutes of the Nineteenth Annual Meeting," *Proc APhA* 19 (1871): 122.

15. E[dward] Parrish, "On the Relations of the Several Classes of Druggists and Pharmacists to the Colleges of Pharmacy," *AJP* 43 (1871): 481.

16. [Procter], "Elias Durand [obit.]," *AJP* 45 (1873): 513.

17. Ibid., 514.

18. "Minutes of the Philadelphia College of Pharmacy," *AJP* 45 (1873): 508–17, esp. 513–14; Sonnedecker, "Pharmaceutical Education," 448; Procter apparently did this with some difficulty and never learned German (Procter to Albert Ebert, January 10, 1869, in A2: Procter, Kremers Reference Files, F. B. Power Pharmaceutical Library, School of Pharmacy, University of Wisconsin-Madison [hereafter cited as KRF]).

19. For more detail concerning pharmaceutical education, see chap. 6. See also Fisher, "A Brief Sketch of the Progress and Present State of Pharmacy," 275f.; Sonnedecker, "Pharmaceutical Education," 28, 268, 275, 374.

20. Sonnedecker, "Pharmaceutical Education," 374.

21. Fisher, "A Brief Sketch of the Progress and Present State of Pharmacy," 275.

22. Sonnedecker, "Pharmaceutical Education," 433–34; Joseph England, ed., *The First Century of the Philadelphia College of Pharmacy, 1821–1921* (Philadelphia: Philadelphia College of Pharmacy, 1922), 400–401. Carson (1808–77) is particularly remembered for his history of the Medical Department of the University of Pennsylvania published in 1869.

23. England, *First Century,* 399–401; Sonnedecker, *Kremers and Urdang's History of Pharmacy,* 264–65.

24. "Minute Book of the Trustees of the Philada. College of Apothecaries [Pharmacy] [Vol. 1, 1821–40]," ms. bound volume, rare book collection, Philadelphia College of Pharmacy and Science, 333.

25. Bache, "Address," 96–100.

26. Carson, "Address," 97–101.

27. [Bullock], "Memoir of William Procter," 519.

28. This is quoted by Bullock, who apparently had unrestricted access to Procter's personal papers for the preparation of his memoir. Procter and Bullock had worked together on the College's committee on deceased members and may have agreed to allow full access when either died ([Bullock], "Memoir of William Procter," 520).

29. Joseph Remington wrote the best of these "pen portraits" of Procter (see Remington, "Procter Reminiscences," 549–55).

30. Edwin M. Boring, "Tribute to the Memory of Professor Procter," *AJP* 76 (1905): 41; Albert E. Ebert, "The Father of American Pharmacy," *Proc APhA* 50 (1902): 153–56; P. W. B[edford], "Obituary Record of Professor William Procter, Jr.," *Drugg Circ* 18 (1874): 68; John F Hancock, "William Procter, Jr., the Father of American Pharmacy," *AJP* 77 (1905): 15–16; F Hoffmann, James Shinn, and George W. Sloan, "Recollections and Reminiscences of Prof. Wm. Procter, Jr.," *AJP* 72 (1900): 485; Remington, "Procter Reminiscences," 550, 557.

31. Evan T. Ellis, "Tribute to the Memory of Professor Procter," *AJP* 76 (1905): 41. The Ellis pharmacy acted as the business office of the *American Journal of Pharmacy* during the 1850s and 1860s.

32. "Minutes of the Philadelphia College of Pharmacy," *AJP* 46 (1874): 132.

33. Hoffman et al., "Recollections," 485; Procter to Ebert, March 5, 1870, A2: Procter, KRF; cf. "Minutes of the Committee on Historical Pharmacy," *Proc APhA* 52 (1904): 432f.

34. Remington, "A Memorial of Procter," 27–28; Remington, "Procter Reminiscences," 554, 555.

35. [Bullock], "Memoir of William Procter," 529; Samuel M. Colcord, "Pharmaceutical Colleges and Associations: Massachusetts College of Pharmacy," *AJP* 46 (1874): 136.

36. John F Hancock, "Pharmaceutical Colleges and Associations: Maryland College of Pharmacy," *AJP* 46 (1874): 138; Thomas S. Wiegand, "Tribute to the Memory of Professor Procter," *AJP* 77 (1905): 43.

37. [Procter], "Editorial Department," *AJP* 31 (1859): 584.

38. Ibid., *AJP* 32 (1860): 571.

39. Edward Pessen, *Jacksonian America: Society, Personality, and Politics* (Homewood, Ill.: Dorsey Press, 1978), 17–20; Arthur Charles Cole, *The Irrepressible Conflict, 1850–1865* (New York: Macmillan, 1934), 160.

40. [Procter], "Editorial Department," *AJP* 32 (1860): 572.

41. "First Session—Monday Afternoon, Sept. 16, 1901," *Proc APhA* 49 (1901): 24.

42. Joseph Remington, "Tributes to the Memory of Professor Procter," *AJP* 77 (1905): 37; see also Remington, "Procter Reminiscences," 550.

43. [Procter], "Editorial Department," *AJP* 37 (1865): 77. See also review of Dunglison's *New Remedies* in ibid., *AJP* 29 (1857): 88.
44. Procter to Ebert, August 20, 1872, A2: Procter, KRF
45. Remington, "Procter Reminiscences," 556, 560.
46. [Bullock], "Memoir of William Procter," 526; Remington, "Procter Reminiscences," 554; Lewis Saum, *The Popular Mood of Pre-Civil War America* (Westport, Conn.: Greenwood Press, 1980), 181f.; Robert Lacour-Gayet, *Everyday Life in the United States before the Civil War, 1830–1860,* trans. Mary Ilford (New York: Frederick Ungar, 1969), 101.
47. [Bullock], "Memoir of William Procter," 519–20.
48. [Procter], "Editorial Department," *AJP* 23 (1851): 386–88. Procter was a friend of Henry A. Tilden, pharmaceutical manufacturer in New Lebanon, New York. Procter also enjoyed visiting the Shaker herb farms in the Lebanon Valley.
49. Although Procter's notebooks apparently did not survive, they are described by Bullock, and their nature is revealed through Procter's published notes. See [Procter], "Editorial Department," *AJP* 39 (1867): 565–78; [Procter], "Notes of Travel in Europe," *AJP* 40 (1868): 17–29; and [Procter], "The Paris Exposition of 1867," *AJP* 41 (1869): 8–13, 108–14.
50. [Procter], "Editorial Department," *AJP* 25 (1853): 371–72. This visit occurred before the fire at the Smithsonian, so no records of this meeting exist in the institution's archive.
51. [Bullock], "Memoir of William Procter," 526. Procter missed the 1867 meeting when he represented the association at the International Pharmaceutical Congress in Paris (see below).
52. [Procter], "Editorial Department," *AJP* 30 (1858): 576, on Procter's visit to Mount Vernon; ibid., *AJP* 37 (1865): 314–15, on his trip to Boston; ibid., *AJP* 38 (1866): 379.
53. Procter, "On a New Variety of Flaxseed," *AJP* 26 (1854): 493–94; Cole, *Irrepressible Conflict,* 18f.
54. [Procter], "Notes of Travel in Europe," *AJP* 40 (1868): 17–18; [Procter], "The Paris Exposition of 1867," *AJP* 41 (1869): 9–10; Procter to Ebert, April 20, 1873, KRF; Ebert, "Father of American Pharmacy," 155f.
55. [Procter], "Editorial Department," *AJP* 39 (1867): 565–66. He later expressed regret about such waywardness (Procter to Ebert, January 10, 1869, KRF).
56. [Procter], "Notes of Travel in Europe," 23–24.
57. [Procter], "The Paris Exposition of 1867," 518ff.; "Minutes of the Philadelphia College of Pharmacy," *AJP* 39 (1867): 563.
58. [Procter], "Notes of Travel in Europe," 19f., 23–28.
59. [Procter], "Botany in Its Bearings on Pharmacy—Botanical Gardens," *AJP* 40 (1868): 235. Daniel Hanbury, a friend and an eminent pharmaceutical botanist, probably guided Procter through Kew Gardens. [Procter], "Notes of Travel in Europe," 18; Sonnedecker, *Kremers and Urdang's History of Pharmacy,* 461.
60. Procter, "Speech Delivered at the International Pharmaceutical Congress, August 21, 1867," translation from the French by George Urdang, A2: Procter, Kremers Reference Files. For text of this speech, see appendix A.
61. "Second Session—Tuesday Morning, May 8, 1900," *Proc APhA* 48 (1900): 30.

62. [Bullock], "Memoir of William Procter," 523; Remington, "Procter Reminiscences," 556.
63. [Procter], "Infusum Juniperi Compositum," *AJP* 25 (1853): 205.
64. Remington, "Procter Reminiscences," 554f.
65. George Kennedy and John Maisch, "Prefatory Notice," *Proc APhA* 22 (1874): 22.
66. The career of Edward Parrish (1822–72) paralleled that of William Procter in several ways: Parrish wrote the second American pharmaceutical textbook after Procter's, held the chair in practical pharmacy at the College soon after Procter retired (see chap. 6), and took part in the founding of the American Pharmaceutical Association in 1852. A close friend of Procter, his name comes up often in the following pages. See George Urdang, "Edward Parrish, A Forgotten Pharmaceutical Reformer," *Am J Pharm Ed* 14 (1950): 223–32.
67. Remington, "Procter Reminiscences," 555.
68. [Bullock], "Memoir of William Procter," 529; "Died . . . Procter," *Friends' Intelligencer* 30 (1874): 825.
69. Remington, "Procter Reminiscences," 558; "Died . . . Procter," 825. Spruce Street was a Hicksite meeting.
70. Procter, "Valedictory Address to the Graduates of the Philadelphia College of Pharmacy, Delivered March 11th, 1858, at the Musical Fund Hall," *AJP* 30 (1858): 199; [Bullock], "Memoir of William Procter," 529; Benjamin, *Philadelphia Quakers,* 13ff. For example, see [Procter], "Editorial Department," *AJP* 26 (1854): 569–74.
71. Often written hastily by Procter to fit them into the next number of the *Journal,* these obituaries reveal as much about Procter as any of his published writings.
72. [Procter], "Editorial Department," *AJP* 25 (1853): 287–88; "[Minutes of the Second Annual Meeting]," *Proc APhA* 2 (1853): 19.
73. Sonnedecker, *Kremer and Urdang's History of Pharmacy,* 487.
74. It probably is not a coincidence that Wallace Procter, William's son, wrote his College thesis on the medicinal properties of species closely related to the one analyzed by Stephen thirty-eight years before (Wallace Procter, "On the Fruit of Magnolia Tripetala," *AJP* 44 [1872]: 145f.).
75. Procter, "Pharmaceutical Notices," *AJP* 13 (1841): 184. Procter's respect for Durand is shown in a note that he added to the report of the Philadelphia College of Pharmacy's committee on the *Pharmacopoeia,* where he defends Durand's chronic absences. [Procter], "Journal of the Committee on the Revision of the U.S. Pharmacopoeia, of the Philadelphia College of Pharmacy," manuscript book, in dean's office of Philadelphia College of Pharmacy and Science, 51.
76. [Procter], "Necrology [Augustine Duhamel]," *AJP* 18 (1846): 316. Duhamel was instrumental in introducing displacement (percolation), which eventually became a standard extraction method.
77. [Procter], "A Concise Historical Sketch of the Progress of Pharmacy in Great Britain," *AJP* 20 (1848): 265.
78. [Procter], "Editorial Department," *AJP* 23 (1851): 188–89.
79. Ibid., *AJP* 31 (1859): 490.

80. Ibid., *AJP* 29 (1857): 283; *AJP* 32 (1860): 192; *AJP* 34 (1862): 491; *AJP* 40 (1868): 94; Daniel Hanbury, "On the Manufacture of Balsam of Peru," *AJP* 36 (1864): 145; Procter to Ebert, January 3, 1868, February 28, 1869, A2: Procter, KRF
81. Curti, *American Thought,* 311, 366–68.
82. Ibid., 360.
83. Russel B. Nye, *Society and Culture in America, 1830–1860* (New York: Harper and Row, 1974), 20, 29–30, 34f.; Miller, *Modern America,* 35; Arthur Alphonse Ekirch, *The Idea of Progress in America, 1815–1860* (New York: Peter Smith, 1951), 118, 120.
84. [Procter], "Editorial Department," *AJP* 26 (1854): 92–93; Miller, *Modern America,* x.
85. Ekirch, *Idea of Progress,* 36.
86. Procter, "On the Decomposing Power of Water at High Temperatures, in a Scientific and Industrial Point of View, as Developed by R. A. Tighlman," *AJP* 20 (1848): 184.
87. [Procter], "On the Production of Wine Brandy, and Tartar in the Valley of the Ohio," *AJP* 26 (1854): 411–14; [Procter], "Editorial Department," *AJP* 25 (1853): 381; Oscar Oldberg in Remington, "Procter Reminiscences," 556f.
88. Curti, *American Thought,* 381; [Procter], "Editorial Department," *AJP* 26 (1854): 571; Lacour-Gayet, *Everyday Life,* 264.
89. Edward Parrish, "American Pharmacy," *AJP* 26 (1854): 117–18; [Procter], "Chemical Technology or Chemistry Applied to the Arts and Manufactures by Dr. F Knapp [review]," *AJP* 20 (1848): 280.
90. Curti, *American Thought,* 346–49.
91. Procter, "Lecture, Introductory to the Course on Practical Pharmacy," 244; Remington, "Procter Reminiscences," 555.
92. [Procter], "Historical Sketch," 265.
93. [Procter], "Editorial Department," *AJP* 36 (1864): 287; *AJP* 29 (1857): 478; Procter footnote in George Kemp, "On a Direct Method of Obtaining Iodine from Certain Species of Sea-weed," *AJP* 22 (1850): 348. Cf. Ekirch, *Idea of Progress,* 36.
94. Procter to Ebert, August 20, 1872, A2: Procter, KRF; "First Session," 24; Procter to Ebert, January 9, 1870, A2: Procter, KRF; Remington, "Procter Reminiscences," 550.
95. A. G. DuMez, "Procter, William," in *Dictionary of American Biography,* vol. 8 (1935), 242; [Bullock], "Memoir of William Procter," 525.
96. [Bullock], "Memoir of William Procter," 526; B[edford], "Obituary," 68.
97. Remington, "A Memorial," 27.
98. "Will of Catherine Procter, decd," book 268, no. 1292, Philadelphia City Hall, Registry of Wills, 1905, n.p.
99. England, *First Century,* 387; "Pharmaceutical Colleges and Associations: Philadelphia College of Pharmacy," *AJP* 44 (1872): 180–81.
100. To his brother Stephen, he gave $500 and all his medical books; to niece Elizabeth Goldsmith, $500; to the trustees of the College, $500 in trust for an annual award to the best graduate; and $100 per year for life to the father of his first wife, Amos Bullock. The rest of his property was to be divided

among his family, half to Catherine and the other half divided between Wallace and Mary (see note 101 below).

101. "Will [and associated documents] of William Procter, Jr., decd," book 80, no. 121, p. 418, Philadelphia City Hall, Registry of Wills, 1874, 4. After Procter's death in 1874, Preston and Wallace Procter did go into partnership together until 1890 (England, *First Century,* 387).

102. "Died: Procter," (Philadelphia) *Public Ledger,* vol. 76, no. 120 (February 11, 1874), 2.

103. "Registration of Deaths in the City of Philadelphia, 1874," p. 50, in Philadelphia City Archive, City Hall Annex.

104. "William Procter, Jun.," *Pharmaceutical Journal and Transactions* 33 (1874): 707.

105. "Obituary: Prof. Wm. Proctor [*sic*], Jr.," *The Pharmacist* 7 (1874): 94.

106. "William Procter, Jun.," 707.

2. William Procter as Practicing Pharmacist

1. [Bullock], "Memoir of William Procter," 520–21.

2. Fisher, "A Brief Sketch of the Progress and Present State of Pharmacy," 271; *McElroy's Philadelphia Directory for 1845* (Philadelphia: Biddle, 1845), 290, 411; Galen, "Philadelphia Correspondence of the Druggist's Circular," *Drugg Circ* 1 (1857): 137.

3. Curti, *American Thought,* 292ff.; Fisher, "A Brief Sketch of the Progress and Present State of Pharmacy," 271; David H. Donald, *Liberty and Union* (Boston: Little, Brown, 1978), 24–25, 183.

4. Fisher, "A Brief Sketch of the Progress and Present State of Pharmacy," 272f.; Sonnedecker, *Kremers and Urdang's History of Pharmacy,* 290ff.; George Urdang, "The Part of Doctors of Medicine in Pharmaceutical Education," *Am J Pharm Ed* 14 (1950): 548.

5. John Faber, "On Pharmacy in the United States," *AJP* 41 (1869): 402–3; Sonnedecker, "Pharmaceutical Education," 516f.; [Procter], "Editorial Department," *AJP* 28 (1856): 480; [L. V. Newton], "What is a Drug Store Proper?" *Drugg Circ* 3 (1859): 278; Procter, "Lecture, Introductory to the Course on Practical Pharmacy," 252; Sonnedecker, *Kremers and Urdang's History of Pharmacy,* 290f.

6. Rowland Berthoff, "Independence and Enterprise: Small Business in the American Dream," in *Small Business in American Life,* ed. Stuart W. Bruchey (New York: Columbia Univ. Press, 1980), 31–34.

7. Bache, "Address," 101; Parrish, "American Pharmacy," 212f.; John M. Maisch, "Physicians and Pharmaceutists, and Their Relations," *AJP* 28 (1856): 112; [Procter], "Editorial Department," *AJP* 23 (1851): 186; Procter, "Thoughts on 'Manufacturing Pharmacy' in Its Bearing on the Practice of Pharmacy, and the Character and Qualifications of Pharmaceutists," *AJP* 30 (1858): 516.

8. [Bullock], "Memoir of William Procter," 521.

9. Procter, "Observations on the Volatile Oil of Betula Lenta, and on Gaultherin, a Substance, which by its Decomposition, Yields that Oil," *AJP*

15 (1843): 241–50; Augustine Duhamel and William Procter, Jr., "Report on a Root Found in Senega," *AJP* 15 (1843): 256–57; Procter, "Observations on Modardin, a Peculiar Crystalline Substance, Derived from the Volatile Oil of Monarda Punctata," *AJP* 17 (1845): 86–89; Procter, "Observations on a New Displacement Apparatus, Invented and Patented by Charles Augustus Smith, of Cincinnati," *AJP* 18 (1846): 98–103.

10. "Minutes of the Philadelphia College of Pharmacy," *AJP* 18 (1846): 147; [Bullock], "Memoir of William Procter," 522.

11. For his first year of teaching, 1847–48, Procter received about $200 from the sale of student fee tickets. This is estimated from the $75 brought into the College from the 12.5 percent tax it placed on each of its three professors ("Minutes of the Philadelphia College of Pharmacy," *AJP* 19 [1847]: 155).

12. This stipend began at about $150 and rose to $400 by the end of Procter's editorship.

13. Procter to Ebert, January 9, 1870, A2: Procter, KRF

14. This was a good day; usually he received closer to one dollar (Prescription book of William Procter, Jr., January–March 1847, Archives, American Pharmaceutical Association, Washington, D.C.).

15. There is no price marked on the prescription for this small item; Procter may have given it to Mayer gratis (Procter prescription book, no. 323).

16. Procter prescription book, no. 324. This prescription has the name of the patient on the top, a fairly rare occurrence in Procter's book; only about one prescription in ten has any name on it. None of the 414 prescriptions examined contained all the components required today by law: patient name, date, ingredients, directions for preparation and use, and doctor's signature.

17. The prices seem modest, for when Procter collected his fifty cents, the wholesale price index (all commodities) stood at 64.9 as estimated by the Bureau of Labor Statistics. A century later, the average prescription price was between $1.40 and $1.47, when the index stood at 152.1 (*Historical Statistics of the United States, 1789–1945* [Washington, D.C.: GPO, 1949], 234; and *"Lilly Digest . . .* Operating Statements of 1947," *Journal of the American Pharmaceutical Association* [Prac. Pharm. Ed.], 9 [1948]: 681).

18. Procter prescription book, no. 325. Comment on powder papers from Francis Mohr, Theophilus Redwood, and William Procter, Jr., *Practical Pharmacy* (Philadelphia: Lea and Blanchard, 1849), 492. By adding potassium nitrate to the Dover's powder of the 1840s, one could approximate the formula of the original preparation of Dover (George B. Wood and Franklin Bache, *The Dispensatory of the United States,* 13th ed. [Philadelphia: J. B. Lippincott, 1870], 1368).

19. Procter prescription book, no. 326; Mohr, Redwood, and Procter, *Practical Pharmacy,* 474f.

20. Procter prescription book, no. 327. Procter continued to dispense polypharmaceutic tonics throughout his career. See Procter, "Pharmaceutical Notices," *AJP* 41 (1869): 390, for a description of a tonic with twelve ingredients, including opium, cannabis, belladonna, cinchona, and strychnine.

21. Procter prescription book, no. 328. Wood and Bache, *Dispensatory* (13th ed.), 229n. February 28, 1847, fell on a Sunday and Procter filled no prescriptions that day.

22. Cole, *Irrepressible Conflict,* 150. Skilled mechanics, men comparable in social rank to pharmacists, earned about two dollars a day.
23. Procter prescription book, January 1 to March 15, 1847. He probably obtained these ingredients from a local wholesale druggist, such as Powers and Weightman.
24. John Harley Warner, *The Therapeutic Perspective: Medical Practice, Knowledge, and Identity in America, 1820–1885* (Cambridge, Mass.: Harvard Univ. Press, 1986), 4–7, 91–101.
25. Ibid., 98.
26. Charles Rosenberg, in *The Therapeutic Revolution: Essays in the Social History of American Medicine* (Philadelphia: Univ. of Pennsylvania Press, 1979), 9.
27. David L. Cowen, Louis D. King, and Nicholas G. Lordi, "Nineteenth Century Drug Therapy: Computer Analysis of the 1854 Prescription File of a Burlington [New Jersey] Pharmacy," *Journal of the Medical Society of New Jersey* 78 (1981): 760.
28. Ibid., 758–61.
29. Procter's additions are marked in brackets. Many of these supplements seem to be based on personal experience.
30. Mohr, Redwood, and Procter, *Practical Pharmacy,* 22.
31. For example, tincture of kino (Procter prescription book, no. 66) gelatinized and lost its astringency over time. See Mohr, Redwood, and Procter, *Practical Pharmacy,* 278.
32. Ibid., 27–28.
33. Ibid., 470.
34. Ibid., 45.
35. Ibid., 94–102.
36. Ibid., 152.
37. Ibid., 163.
38. Ibid., 278.
39. Ibid., 282f., 412–16.
40. Ibid., 477. Later Procter would suggest that pharmacists add a volume on therapeutics to their libraries so that they would better understand what they dispensed ([Procter], "Editorial Department," *AJP* 39 [1867]: 287).
41. [Procter], "Editorial Department," *AJP* 29 (1857): 86.
42. [Procter], "Remarks on a Carbonic Acid Water Apparatus," *AJP* 28 (1856): 103; Procter, "Eau de Vichy: Artificial Vichy Water," *AJP* 29 (1857): 405f.; Remington, "Tributes," 37; Remington, "Procter Reminiscences," 550; [Bullock], "Memoir of William Procter," 527–28. Joseph P. Remington contended that Procter could have doubled his business if he had moved from South Philadelphia, but that he was too timid and conservative to do it (Remington, "A Memorial," 23).
43. Remington, "Procter Reminiscences," 550, on the increase of business. In the nineteenth and early twentieth centuries, employee pharmacists were referred to as "clerks." As such Preston had responsibility for the pharmacy in the absence of the owner.
44. "Will [and associated documents] of William Procter, Jr., decd," n.p. In his will of 1867, Procter stipulated that Preston be allowed to buy a full partnership in the family business.

45. [Procter], "Editorial Department," *AJP* 23 (1851): 92; *AJP* 38 (1866): 187. See also Procter, "Valedictory Address," 195.

46. For example, making flavoring extracts for both medicinals and foods ([Procter], "On Flavoring Extracts," *AJP* 38 [1866]: 294ff.).

47. Procter, "On a Variety of Cinchona Bark," *AJP* 19 (1847): 178–82; Procter, "Note on Adulterated Powdered Socotrine Aloes," *AJP* 25 (1853): 99; [Procter], "Editorial Department," *AJP* 42 (1870): 381.

48. Procter, "Lecture, Introductory to the Course on Practical Pharmacy," 249.

49. Procter, "On Sophistications of Sulphate of Quinia and Blue Mass," *AJP* 18 (1846): 267; [Procter], "Editorial Department," *AJP* 23 (1851): 92; Procter, "Remarks on Gum Mezquite," *AJP* 27 (1855): 14–17; George Urdang, "The Reagent Bottles of William Procter, Jr.," *Pharmaceutical Archives* 14 (1943): 45.

50. John M. Maisch, "On the Chemical Constituents of Coca Leaves," *AJP* 33 (1861): 497.

51. J. M. Wallace, "Antidote to Corrosive Sublimate," *AJP* 15 (1843): 333; "Death of Professor William Proctor [*sic*], Jun.," *The Chemist and Druggist* 15 (1874): 77; Hancock, "William Procter, Jr., the Father of American Pharmacy," 18; Wiegand, "Tribute to the Memory of Professor Procter," 43.

52. Procter, "Note on Cucumber Ointment," *AJP* 19 (1847): 16–17; [Procter], "Description of Some Pharmaceutical Apparatus," *AJP* 25 (1853): 27–31; [Procter], "On Species St. Germain," *AJP* 32 (1860): 312; "Minutes of the Eighteenth Annual Meeting," *Proc APhA* 18 (1870): 85.

53. "Minutes of the Pharmaceutical Meetings," *AJP* 16 (1844): 71f.; Procter, "Pharmaceutical Notices," *AJP* 19 (1847): 259ff.

54. Procter, "Note on Cucumber Ointment," 16–17.

55. Procter, "Lecture, Introductory to the Course on Practical Pharmacy," 241.

56. Ibid., 243.

57. Procter, "Valedictory Address," 193.

58. [Procter], "Editorial Department," *AJP* 30 (1858): 579.

59. Ibid., *AJP* 35 (1863): 576. For Procter, professionalism in pharmacy meant the public recognition of the pharmacist's special knowledge and ability, combined with the pharmacist's self-awareness of his ethical and scientific responsibilities. Cf. *American Dictionary of the English Language,* 1854, s.v., "profession."

60. This is what Polanyi called the "double movement" of the nineteenth century; cf. Starr, *The Social Transformation of American Medicine,* 61, where Starr applies the concept to nineteenth-century American medical practice (Karl Polanyi, *The Great Transformation* [Boston: Beacon Press, 1967], 130ff.).

61. Procter, "Lecture, Introductory to the Course on Practical Pharmacy," 251; Procter, "Remarks on the Revision of the Pharmacopoeia," *AJP* 18 (1846): 2–3; [Procter], "Editorial Department," *AJP* 26 (1854): 90; *AJP* 29 (1857): 86.

62. [Procter], "Editorial Department," *AJP* 26 (1854): 477. See appendix A for Procter's comments in Paris. When the British in 1868 set up a compromise system similar to what Procter advocated, he reacted favorably ([Procter], "On the New Act of Parliament Relative to Pharmacy and the Sale of Poisons," *AJP* 40 [1868]: 397–400).

63. [Procter], "Editorial Department," *AJP* 23 (1851): 189–90; *AJP* 29 (1857): 186f.; *AJP* 32 (1860): 91.

64. Ibid., *AJP* 35 (1863): 383. For a full description of the contemporary methods to detect arsenic in the remains of a victim, see Augustine Duhamel, "Memoir upon the Detection of Arsenic Forming a Toxicological Compendium, as far as regards this Substance," *AJP* 12 (1840): 279–97.

65. [Procter], "Editorial Department," *AJP* 29 (1857): 186f.; *AJP* 32 (1860): 376f.; *AJP* 30 (1858): 470. Procter did not stock arsenic for over-the-counter sale, to avoid discriminating his trustworthy customers from those who were not. He advised others to follow his lead.

66. Ibid., *AJP* 41 (1869): 83–85.

67. Ibid., *AJP* 32 (1860): 91. Procter also provides his own detailed tips on how to prevent such occurrences (pp. 90–91).

68. Ibid., *AJP* 41 (1869): 186; *AJP* 42 (1870): 182; England, *First Century*, 130.

69. Sonnedecker, *Kremers and Urdang's History of Pharmacy*, 216.

70. [Procter], "Editorial Department," *AJP* 40 (1868): 91. Some of these same sentiments were voiced by Oscar Oldberg, an emigrant from Sweden, where the number of pharmacies was limited; Oscar Oldberg, "The Drug Business in Sweden," *AJP* 42 (1870): 29.

71. [Procter], "Editorial Department," *AJP* 40 (1868): 470; *AJP* 41 (1869): 83, 87, 182; [Procter], "Pharmaceutical Titles," *AJP* 43 (1871): 157–58; [Procter], "Editorial Department," *AJP* 43 (1871): 187.

72. Maisch, "Physicians and Pharmaceutists," 204; [Procter], "Editorial Department," *AJP* 40 (1868): 187.

73. [L. V. Newton], "Kings County Physicians against the Apothecaries," *Drugg Circ* 7 (1863): 141; Th. Schumann, "The Relation between the Physician and Pharmaceutist," *AJP* 44 (1872): 403–4; Sonnedecker, "Pharmaceutical Education," 516–17.

74. Procter, "Lecture, Introductory to the Course on Practical Pharmacy," 253; [Procter], "Editorial Department," *AJP* 26 (1854): 381.

75. Mohr, Redwood, and Procter, *Practical Pharmacy*, 477; [Procter], "Editorial Department," *AJP* 23 (1851): 286–88, 385–86. It is a curious reversal that in the late twentieth century prescriptions are written commonly with the ingredients in English and the directions in abbreviated Latin.

76. [Procter], "Editorial Department," *AJP* 25 (1853): 475–76. Procter generally lumped together sectarian and incompetent, "regular" physicians ([Procter], "Eclectic Pharmacy," 108–9). The sectarian battles that tore apart American medicine in the nineteenth century bypassed pharmacy almost completely.

77. [Procter], "Editorial Department," *AJP* 25 (1853): 476; *AJP* 31 (1859): 90.

78. Ibid., *AJP* 24 (1852): 386; *AJP* 25 (1853): 476; "Minutes of the Sixth Annual Meeting," *Proc APhA* 6 (1857): 29–30. Often considered a rural phenomenon, doctor's shops were common in mid-nineteenth-century American cities as well. In 1853 they made up about one-fifth of all retail druggists in Philadelphia ("Editorial Department," *AJP* 26 [1854]: 383–84).

79. England, *First Century*, 327; Jonathan Michael Liebenau, "Medical Science and Medical Industry, 1890–1929: A Study of Pharmaceutical Manufacturing in Philadelphia" (Ph.D. diss., Univ. of Pennsylvania, 1981), 47–54.

80. Edward R. Squibb, "The Manufacture, Impurities, and Tests of Chloroform," *AJP* 29 (1857): 431–32. In the late 1850s Squibb was a small-scale

manufacturer and specialized in chemicals difficult to make in the shop, such as ether and chloroform.

81. Procter, "Manufacturing Pharmacy," 516–17. For example, chloric ether (Procter, "Note on Commercial Chloric Ether," *AJP* 31 [1859]: 553) and cathartic pills ([Procter], "Editorial Department," *AJP* 31 [1859]: 87).

82. See chapter 3 on percolation and fluid extracts. [Procter], "Editorial Department," *AJP* 38 (1866): 287; James W. Mill, "Practical Knowledge," *AJP* 32 (1860): 12; Wm. L. Turner, "The Need of Practical Information in Our Pharmaceutical Publications," *AJP* 44 (1872): 533.

83. [Procter], "Editorial Department," *AJP* 38 (1866): 287.

84. Ibid., *AJP* 41 (1869): 81–82.

85. Ten years before Procter had composed a similar warning about nostrums; see Procter, "Manufacturing Pharmacy," 516. Parrish shared his views; see Parrish, "Several Classes of Druggists and Pharmacists," 483.

86. [Procter], "Editorial Department," *AJP* 37 (1865): 315–16. The stamp tax, mentioned above, was placed on nostrums, which almost all pharmacies sold.

87. Ibid., 314; *AJP* 42 (1870): 183.

88. Ibid., *AJP* 37 (1865): 314ff.; *AJP* 38 (1866): 188.

89. Edward Parrish, "A Discourse on Titles, etc.," *AJP* 39 (1867): 241. Although pharmacies like Parrish's model would not appear for several decades, his suggestion that a pharmacy practitioner be called a "pharmacist" instead of apothecary or pharmaceutist was generally adopted.

90. Higby, "Professionalism," 115–24; Sonnedecker, *Kremers and Urdang's History of Pharmacy,* 213–19; see also chap. 7.

91. Edward Parrish, William Procter, Jr., and Ambrose Smith, "Report of the Philadelphia College of Pharmacy . . . regarding the Pharmaceutical Statistics of Pennsylvania," *Proc APhA* 2 (1853): 30–31; [Procter], "Editorial Department," *AJP* 26 (1854): 383–84; James N. Marks, "First Annual Report of the Pharmaceutical Examining Board of Philadelphia," *AJP* 45 (1873): 84, on the rise of competition; Charles T. Carney, "Report on the Revision of the Pharmacopoeia," *Proc APhA* 7 (1858): 178f., on the panic of 1857; Edward Parrish, "What Shall the Boys Do, These Times," *Drugg Circ* 5 (1861): 155, on economic slowdown of 1861; [John M. Maisch], "Editorial Department," *AJP* 43 (1871): 332f., on popularity of elixirs; Medicus, "Elegant Pharmacy," *AJP* 44 (1872): 161–62, on homeopathy's impact on pharmaceutical "elegance."

3. Pharmaceutical Scientist and Technologist

1. Joseph P. Remington, "A Memorial of William Procter, Jr.," a memorial address read at the forty-eighth annual meeting of the American Pharmaceutical Association, pamphlet in A2: Procter, KRF, p. 13, later published in the *Proceedings*.

2. [Procter], "Editorial Department," *AJP* 32 (1860): 89; *AJP* 37 (1865): 314.

3. For example, Procter, "Note on Fluid Extract of Wild Cherry Bark," *AJP* 32 (1860): 399.

4. [Procter], "Editorial Department," *AJP* 35 (1863): 93; *AJP* 25 (1853): 381. Cf.

George H. Daniels, *American Science in the Age of Jackson* (New York: Columbia Univ. Press, 1968), 65.

5. Oscar Oldberg, *A Course of Home Study for Pharmacists* (Chicago: Apothecaries' Company, 1891), 332–33.

6. James C. Ayer, "The Alkaloids and Proximate Principles," *AJP* 25 (1853): 407.

7. Ibid., 407–8; George D. Coggeshall, "An Address Delivered to the Graduates of the New York College of Pharmacy," *AJP* 18 (1846): 246.

8. Sonnedecker, "Pharmaceutical Education," 449.

9. Alex Berman, "The Impact of the Nineteenth Century Botanico-Medical Movement on American Pharmacy and Medicine" (Ph.D. diss., University of Wisconsin, 1954), 325ff.

10. Edward Parrish, "Eclectic Pharmacy," *AJP* 23 (1851): 329–35; [Procter], "Editorial Department," *AJP* 25 (1853): 284; see also Berman, "Botanico-Medical Movement," 294–98.

11. Procter, "On Lobelia Inflata," *AJP* 9 (1837): 98–109.

12. Ibid., 102.

13. Ibid., 106.

14. Ibid., 105.

15. Ibid., 108

16. Procter, "Observations on Lobelia Cardinalis," *AJP* 11 (1839): 280–83.

17. Procter, "On Lobelina, the Active Principle of Lobelia Inflata, and on Some Other Proximate Principles of the Seed of that Plant," *AJP* 13 (1841): 1–10.

18. Remington, "A Memorial," 7.

19. Procter, "Remarks on Some Pharmaceutical Preparations of Lobelia Inflata," *AJP* 14 (1842): 108–9.

20. Ibid., 108.

21. Procter, "Observations on the Volatile Oil of Gaultheria Procumbens, Proving it to be a Hydracid Analogous to Saliculous Acid," *AJP* 14 (1842): 211–21.

22. Ibid., 221.

23. Procter, "Observations on the Volatile Oil of Betula Lenta, and on Gaultherin," 241–50; E. Gildemeister and F. Hoffmann, *The Volatile Oils,* trans. Edward Kremers (Milwaukee: Pharmaceutical Review Publishing, 1900), 1:118.

24. Procter, "Observations on Monardin," 86–89.

25. Gildemeister and Hoffmann, *The Volatile Oils,* 3:463. For example, see *The Merck Index,* ed. Martha Windholz (Rahway, N.J.: Merck, 1983), 925.

26. Milton Wruble, *Studies in Percolation* (Reprint; Philadelphia: American Journal of Pharmacy, 1933), 5–17.

27. Augustine Duhamel, "Boullay's Filter and System of Displacement, with Observations Drawn from Experience," *AJP* 10 (1838): 1–17.

28. Augustine Duhamel and William Procter, Jr., "Observations on the Method of Displacement," *AJP* 11 (1839): 189–201.

29. Ibid., 189–90.

30. Ibid., 190–95, 199.

31. W. Procter, Jr. and J. C. Turnpenny, "Remarks on Syrup of Wild Cherry Bark, and on Syrup of Valerian," *AJP* 14 (1842): 27–28.

32. Procter, "Remarks on the Revision of the Pharmacopoeia," 9.
33. Edward R. Squibb, "The Process of Percolation," *AJP* 30 (1858): 97.
34. Wood and Bache, *Dispensatory,* 13th ed., 1153.
35. Ibid.; see also W. B. Chapman, "On the Uncertainty of the Composition of Pharmaceutical Preparations, and the Most Eligible Form of Medicines for Administration," *AJP* 26 (1854): 157.
36. Wood and Bache, *Dispensatory,* 13th ed., 1152.
37. Procter, "Formulae for the Fluid Extracts in Reference to Their More General Adoption in the Next Pharmacopoeia," *AJP* 31 (1859): 530–31.
38. Procter, "On the Fluid Extracts of Rhubarb and Valerian," *AJP* 19 (1847): 182; Procter, "Remarks on Fluid Extracts of Cinchona," *AJP* 23 (1851): 218–19; Procter, "On the Pharmacy of Cimicifuga," *AJP* 26 (1854): 106–7; [Procter], "Pharmaceutical Notices," *AJP* 29 (1857): 108.
39. Procter, "Formulae for Fluid Extracts," 530–49.
40. "Statement of Prof. Joseph P. Remington of the Philadelphia College of Pharmacy, Philadelphia, Pa.," in *Hearings before the Committee on the Library, H.R. 11076, February 18, 1916* (Washington, D.C.: GPO, 1916), 18.
41. *The Pharmacopoeia of the United States of America,* 1863 ed., 162–80. The revision committee also adopted a new class of preparations suggested by Procter—the oleoresins.
42. "Minutes of the Eighth Annual Meeting," *Proc APhA* 8 (1859): 34; "The Eighth Annual Meeting of the American Pharmaceutical Association," *Drugg Circ* 3 (1859): 223. See also Joseph A. Heintzelman, "Syrupus Stillingiae Compositus," *AJP* 34 (1862): 208.
43. Procter, "Pharmaceutical Notices," *AJP* 34 (1862): 136–37.
44. Procter, "On Fluid Extracts," *Proc APhA* 11 (1863): 226–27.
45. [Procter], "The Officinal Fluid Extracts," *AJP* 37 (1865): 181; "Editorial Department," *AJP* 41 (1869): 82.
46. C. Lewis Diehl, "On Fractional Percolation," *AJP* 41 (1869): 337; Samuel Campbell, "On a New and Simple Process for Fluid Extracts," *AJP* 41 (1869): 385–86; E. H. Sargent, "On Fluid Extracts," *AJP* 42 (1870): 337–40; James W. Mill, "The Strength of Fluid Extracts," *AJP* 43 (1871): 17–19.
47. Mill, "Fluid Extracts," 23–24; Sargent, "Fluid Extracts," 337.
48. [Procter], "Vacuum Maceration: Duffield's Process for Fluid Extracts," *AJP* 41 (1869): 4.
49. Joseph P. Remington, *The Practice of Pharmacy,* 2d ed. (Philadelphia: J. B. Lippincott, 1893), 1006–8, 1016–52.
50. John M. Maisch, "On the Use of Our Indigenous Plants," *Proc APhA* 6 (1857): 156; Remington, "A Memorial" (pamphlet), 7–9; Wiegand, "Tribute," 43; "Minutes of the Twentieth Annual Meeting," *Proc APhA* 20 (1872): 63.
51. Procter, "Observations on the Carbonate and Protomuriate of Iron," *AJP* 10 (1838): 272–75.
52. Procter, "On the Tartrate of Iron and Ammonia," *AJP* 12 (1840): 275; Wood and Bache, *Dispensatory,* 13th ed., 1174.
53. Robert P. Hudson, "The Biography of Disease: Lessons from Chlorosis," *Bulletin of the History of Medicine* 51 (1977): 449; Wood and Bache, *Dispensatory,* 13th ed., 1170f.
54. [Bullock], "Memoir of William Procter," 527; Procter, "Remarks on the

Reduction of Iron by Hydrogen," *AJP* 19 (1847): 11ff.; Procter, "Pulvis Ferri: Iron by Hydrogen," *AJP* 26 (1854): 217–18. T. A. Quevenne, a French physician, studied its physiologic properties extensively during the 1840s and 1850s and popularized its medicinal use (Wood and Bache, *Dispensatory,* 13th ed., 404, 1196).

55. Procter, "Observations on Hydrated Peroxide of Iron," *AJP* 14 (1842): 30–31.

56. Wallace, "Antidote to Corrosive Sublimate," 332–33; Mohr, Redwood, and Procter, *Practical Pharmacy,* 379. Ten years later, when a life was lost because an incompetent pharmacist failed to make the ferrous oxide properly, Procter repeated his recipe (Procter, "Remarks on the Preparation of Hydrated Sesquioxide of Iron as an Antidote," *AJP* 25 [1853]: 104–7).

57. Wood and Bache, *Dispensatory,* 13th ed., 29–31, 1184; Procter, "On Liquor Ferri Nitratis U.S.P., 1850; and on a Formula for Syrup of Proto-Nitrate of Iron," *AJP* 23 (1851): 312–15; Procter, "On a New Process for Making Liquor Ferri Nitratis," *AJP* 29 (1857): 306–7; [Procter], "Note on 'Alcoholized Iron,'" *AJP* 39 (1867): 11–12; Professor Attfield, "Observations on Ferric Hydrate," *AJP* 40 (1868): 455.

58. Mohr, Redwood, and Procter, *Practical Pharmacy,* 101–2, 443–44, 272–73, 391–97, 345–46.

59. Procter, "On a Still for Apothecaries," *Proc APhA* 11 (1863): 207–8; Charles O. Curtman, "Pharmaceutical Still," *AJP* 41 (1869): 197–98; Edward Parrish, "Illustrations of Some Pharmaceutical Processes and Apparatus, as Exhibited to the Class in the Philadelphia College of Pharmacy," *AJP* 44 (1872): 5; "Minutes of the Nineteenth Annual Meeting," *Proc APhA* 19 (1871): 117; Remington, *Practice of Pharmacy,* 153.

60. [Procter], "Varieties," *AJP* 29 (1857): 177–85; *AJP* 30 (1858): 274–77, 374–76.

61. [Bullock], "Memoir of William Procter," 519.

62. [Procter], "Sulphocyanide of Mercury: Pharaoh's Serpents. Oleate of Soda and Soap Bubbles," *AJP* 38 (1866): 99–102.

63. Procter, "Lecture, Introductory to the Course on Practical Pharmacy," 242–43. A good example of this type of event was Scheele's discovery of glycerin while making lead plaster.

64. [Procter], "Editorial Department," *AJP* 26 (1854): 92–93, on the telegraph; [Procter], "Notice of M. Carre's Apparatus for Making Ice," *AJP* 42 (1870): 102–6, on Carre's ice-making machine. For examples of Procter's grasp of inorganic chemistry and mineralogy, see Procter, "Description and Analysis of a Mineral," *AJP* 12 (1840): 108–12; and Procter, "On the Decomposing Power of Water at High Temperatures," 184–97.

65. Procter, "Lecture, Introductory to the Course on Practical Pharmacy," 245; [Procter], "Editorial Department," *AJP* 28 (1856): 191.

66. Procter, "Observations on a New Displacement Apparatus," 98–99; Procter, "On Sulphate of Anilin," *AJP* 34 (1862): 295.

67. Remington, "A Memorial," 25; [Procter], "Editorial Department," *AJP* 31 (1859): 288; and *AJP* 32 (1860): 477; "On Some of the Applications of Glycerine," *AJP* 33 (1861): 158–59n.

68. [Procter], "Editorial Department," *AJP* 30 (1858): 475.

69. F. Victor Heydenreich, "On Capsicum Annuum," *AJP* 30 (1858): 296ff.;

Maisch, "Our Indigenous Plants," 156; Remington, "A Memorial" (pamphlet), 7–9; Samuel P. Sadtler, "Influence of Pharmacists on the Development and Advance of Modern Chemistry," *AJP* 93 (1921): 199.

70. [Bullock], "Memoir of William Procter," 516.
71. Ibid., 519–20; "Minutes of the Philadelphia College of Pharmacy," *AJP* 45 (1873): 514.
72. Lacour-Gayet, *Everyday Life,* 101; [Bullock], "Memoir of William Procter," 521.
73. Lacour-Gayet, *Everyday Life,* 115; Benjamin, *The Philadelphia Quakers,* 7; cf. Charles E. Rosenberg, *No Other Gods* (Baltimore: Johns Hopkins Univ. Press, 1976), 137f.; Procter, "On the Decomposing Power of Water at High Temperatures," 184.
74. Israel Grahame (1819–99) was a Baltimore pharmacist and professor at the Maryland College of Pharmacy. He moved to Philadelphia in the 1860s (England, *First Century,* 100–101, 115; "Minutes of the Philadelphia College of Pharmacy," *AJP* 45 (1873): 512f.; Liebenau, "Medical Science and Medical Industry").
75. [Bullock], "Memoir of William Procter," 519.
76. Procter, "Pharmaceutical Notices," *AJP* 13 (1841): 184; [Procter], "Editorial Department," *AJP* 36 (1864): 288. The importance of medical men in the development of the pharmaceutical sciences is discussed in Urdang, "The Part of Doctors of Medicine in Pharmaceutical Education," 546–56.
77. Procter, "Observations on Some of the Camphoriferous Essential Oils," *AJP* 10 (1838): 17–24.
78. "Minutes of the Philadelphia College of Pharmacy," *AJP* 14 (1842): 169–70.
79. "Minutes of the Pharmaceutical Meetings," *AJP* 15 (1843): 234–35, 313–14; *AJP* 16 (1844): 310–13.
80. For example, at the December 4, 1843, meeting, Procter presented preliminary results from his work on *Betula lenta* ("Minutes of the Pharmaceutical Meetings," *AJP* 15 [1843]: 315). Procter helped to revive the tradition of the meetings in the early 1870s ("Minutes of the Pharmaceutical Meetings," *AJP* 43 [1871]: 89f., 134–38).
81. Procter, "Description and Analysis," 108ff.
82. [Procter], "Note on Kavaine and Methysticine," *AJP* 32 (1860): 133.
83. [Procter], "Editorial Department," *AJP* 25 (1853): 371–72. Smithsonian Institution Archives, Record Unit 52, Assistant Secretary, 1850–77, Incoming Correspondence, vol. 6, p. 228, and vol. 8, p. 156; Outgoing Correspondence, vol. 6, p. 396.
84. Procter to Ebert, February 28, 1869, KRF; see also chap. 1.
85. Donald Zochert, "Science and the Common Man in Ante-Bellum America," *Isis* 65 (1974): 448–93; George H. Daniels, et al., *Science In American Society* (New York: Knopf, 1971), 137, 142; Ekirch, *Idea of Progress,* 36.
86. Carl Russell Fish, *The Rise of the Common Man: 1830–1850* (New York: Macmillan, 1927), 91.
87. [Procter], "Editorial Department," *AJP* 24 (1852): 183; *AJP* 36 (1864): 287; [Procter], "Gleanings," *AJP* 29 (1857): 315; [Procter], "Culture of Opium in the United States," *AJP* 41 (1869): 120–21.

88. Procter, "On Gelseminum Sempervirens or Yellow Jassamin," *AJP* 24 (1852): 307–8.

89. Daniels, *American Science,* 3; [Procter], "Editorial Department," *AJP* 33 (1861): 575.

90. [Procter], "Editorial Department," *AJP* 25 (1853): 381, 572.

91. Procter, "Lecture, Introductory to the Course on Practical Pharmacy," 247.

92. [Procter], "Chemical Technology," 280.

93. [Procter], "Editorial Department," *AJP* 39 (1867): 89. Procter's successor as editor, John M. Maisch, received some of the same criticism. See also Turner, "The Need of Practical Information in Our Pharmaceutical Publications," 534.

94. Procter, Eugene Dupuy, and James Cooke, "Report on the Progress of Pharmacy," *Proc APhA* 6 (1857): 50–79. The annual Report on the Progress of Pharmacy provides perhaps the best single English-language source of information on pharmaceutical science in the late nineteenth century.

95. Other Americans whose papers are listed are: Augustine Duhamel (2), Edward Parrish (2), Edward Squibb (3), Joseph Carson (8), and Daniel B. Smith (10). [Royal Society of London], *Catalogue of Scientific Papers . . .* (New York: Johnson Reprint, 1965).

96. Joseph P. Remington, "Edward Robinson Squibb, M.D.," *AJP* 73 (1901): 421–23.

97. Ibid., 424–28.

98. Remington, "Tributes," 37; Sadtler, "Influence of Pharmacists," 199; Ivor Griffith, "A Tribute to William Procter, Father of American Pharmacy," tearsheet from an unknown periodical, A2: Procter, KRE, 50. Procter's influence can be seen in the approach of some of his students; see Albert E. Ebert, "Pharmaceutical Notes," *AJP* 42 (1870): 540.

4. The Pursuit of Drug Quality

1. [Procter, and Joseph Carson], "Editorial Department," *AJP* 22 (1850): 280.

2. Chapman, "On the Uncertainty of the Composition of Pharmaceutical Preparations," 154–55.

3. Procter, "On Sophistications of Sulphate of Quinia and Blue Mass," 257.

4. "The Sponge Divers of Calymnos," *AJP* 38 (1866): 60–61; J. M. Maisch, "On Commercial Belladonna Leaves," *AJP* 34 (1862): 126–27; Procter, "On Sophistications of Sulphate of Quinia and Blue Mass," 257–58; cf. "Report of the Committee on Adulterations and Sophistications of Drugs, Medicines, Chemicals, etc.," *AJP* 23 (1851): 19, where only four of thirteen samples of a crude drug examined were adulterated.

5. "Minutes of the Fifth Annual Meeting," *Proc APhA* 5 (1856): 26.

6. Charles V. Hagner, "On Drug Grinding," *AJP* 23 (1851): 197–206.

7. "Report of the Committee on Adulterations," 19–20.

8. Stricter regulations prevented their sale in Europe; England, *First Century,* 130–31.

9. F. Mahla, "Adulteration of Medicines," *AJP* 40 (1868): 547.

10. [Procter], "On the Importance of a More Conscientious Attention to Unifor-

mity of Strength in the Preparation of Opium," *AJP* 22 (1850): 210; [Procter], "Editorial Department," *AJP* 23 (1851): 291.

11. Procter, "On a Variety of Cinchona Bark," 178–82; Procter, "Note on a Sample of So-Called Opium from Illinois," *AJP* 42 (1870): 68.

12. J. Worth Estes, "John Jones's *Mysteries of Opium Reveal'd* (1701): Key to Historical Opiates," *Journal of the History of Medicine and Allied Sciences* 34 (1979): 208–9. The wide variation in laudanum strength that Estes describes in eighteenth-century England continued on into the mid-nineteenth century in the United States.

13. Squibb, "The Manufacture, Impurities, and Tests of Chloroform," 432.

14. [Procter], "Editorial Department," *AJP* 23 (1851): 92.

15. See Procter, "Note on Adulterated Powdered Socotrine Aloes," 99–101, for a good example of Procter's methods.

16. Procter, "Valedictory Address," 197.

17. In the following discussion, the *Pharmacopoeia of the United States of America* will be sometimes abbreviated *USP,* with revisions indicated by Roman numerals. For example, *"USP II"* refers to the second revision (third edition) of the *Pharmacopoeia.*

18. Sonnedecker, *Kremers and Urdang's History of Pharmacy,* 260–62.

19. Ibid., 263–64.

20. Ibid., 264, 279ff. The *Dispensatory,* or *USD* as it was usually abbreviated, was the most successful book of its type ever published in English, through twenty-six editions from 1833 to 1967.

21. Ibid., 279; [Procter], "Editorial Department," *AJP* 23 (1851): 389.

22. From 1833 to 1879, Lippincott published both the *USD* and the *USP,* probably using revenue from the *Dispensatory* to offset the costs of the *Pharmacopoeia* (Sonnedecker, *Kremers and Urdang's History of Pharmacy,* 279–80). During this period, the *Dispensatory* was more indicative of trends in American therapeutics than the *Pharmacopoeia;* it was revised more often and gave recommendations concerning superior preparations and formulas from other pharmacopoeias. It also reflected current attitudes toward domestic and foreign remedies.

23. England, *First Century,* 120; Urdang, "The Part of Doctors of Medicine in Pharmaceutical Education," 553.

24. Procter, "[Papers Relating to the Revision of the *U.S. Pharmacopoeia,* A.D. 1839]," George Bacon Wood Papers, College of Physicians of Philadelphia, box 3, folder 7; see also anonymous documents in box 3, folder 8.

25. "Minutes of the Committee for Revising and Publishing the U.S. Pharmacopoeia Appointed by the National Medical Convention in Jany. 1840," ms. book in Wood Papers, box 3.

26. Documents Relating to the Pharmacopoeia of the United States of America III [1840]," Wood Papers, box 3, folder 8.

27. "Minutes of the Philadelphia College of Pharmacy," *AJP* 12 (1840): 75–76. See "National Medical Convention," *AJP* 11 (1840): 347–48, for a description of the decision of the physicians to ask the advice of the pharmacists. Hereafter, the revision committee of the Philadelphia College of Pharmacy will be called the "pharmacy committee" to separate it from the actual revision committee of the *USP.*

28. "Journal of the Committee, on the Revision of the U.S. Pharmacopoeia, of the Philadelphia College of Pharmacy, [c. 1840]," ms. book 1f.; "Report of the Committee of Revision of the U.S. Pharmacopoeia of the Philadelphia College of Pharmacy 1840," 2 ms. vols. (xliv + 372 pp., vii–viii), both in dean's office, Philadelphia College of Pharmacy and Science. It should be noted that Procter was appointed secretary of the revision committee of the Philadelphia College of Pharmacy, not the actual revision committee empowered by the Medical Convention. The historical literature has mistakenly credited Procter as the secretary of the revision committee; for example, see Sonnedecker, *Kremer and Urdang's History of Pharmacy,* 481. This probably arose from direct copying from Bullock's memoir, which refers to Procter's service on the College committee as simply the "committee to revise the pharmacopoeia"; [Bullock], "Memoir of William Procter," 530.
29. "Journal of the Committee," 4.
30. "Report of the Committee of Revision on the New Pharmacopoeia, Made to the College, at a Special Meeting, Nov. 6th, 1841," *AJP* 13 (1842): 267–69; hereafter cited as "Report on the New Pharmacopoeia."
31. "Journal of the Committee," 16.
32. Ibid., 30.
33. "Report of the Committee (ms.)," 70–74.
34. "Minutes of the Philadelphia College of Pharmacy," *AJP* 13 (1841): 79.
35. "Journal of the Committee," 47; "Report on the New Pharmacopoeia," 278.
36. "Documents Relating to USP III," n.p. The report was sent to the revision committee of the *Pharmacopoeia* on November 7, 1840. The two-volume work is in Procter's hand and includes a personal note signed by him at the end of the committee minutes.
37. "Minutes of Committee, Jany. 1840," n.p.; *The Pharmacopoeia of the United States of America* (Philadelphia: Grigg and Elliot, 1842), xi–xii, hereafter cited as *USP II.*
38. "Report on the New Pharmacopoeia," 273, 276; "Documents relating to USP III," n.p.
39. *USP II,* 252–57.
40. "Report of the Committee (ms.)," 311–57.
41. This was the first revision that included such tests; ibid., 336. For example, see calomel, *USP II,* 118.
42. "Report on the New Pharmacopoeia," 285.
43. "Report of the Committee (ms.)," vii–viii.
44. "Report on the New Pharmacopoeia," 280–81.
45. Procter, "Remarks on the Revision of the Pharmacopoeia," 2–6. Procter attributed the idea of such a pharmaceutical convention to David Stewart of Baltimore.
46. Procter, "Pharmaceutical Notices," *AJP* 19 (1847): 262ff.
47. "Minutes of the Philadelphia College of Pharmacy," *AJP* 21 (1849): 380–81.
48. "Report of the Committee Appointed by the Philadelphia College of Pharmacy to Revise the Pharmacopoeia of U.S., 1850," ms. book (47 pp.), dean's office, Philadelphia College of Pharmacy and Science, 1; hereafter cited as "Report of the Committee, 1850."
49. Each national convention assembled to revise the *United States Pharmacopoeia*

had been called, prior to 1850, the Medical Convention. After 1850, when organized pharmacy sent representatives, the name was changed to the National Convention to Revise the Pharmacopoeia or simply the Pharmacopoeial Convention.

50. [Minutes of the Committee on Revision of the U.S. Pharmacopoeia of the Philadelphia College of Pharmacy, 1847], mss. bound with Journal of the Committee of the Philadelphia College of Pharmacy [c. 1840], 52–76; "Report of the Committee, 1850," 1.

51. "Minutes of the Philadelphia College of Pharmacy," *AJP* 22 (1850): 93.

52. [Procter with Joseph Carson], "Editorial Department," *AJP* 22 (1850): 283–84.

53. Ibid., 285; *Pharmacopoeia of the United States of America* (Philadelphia: J. B. Lippincott, 1863), ix, hereafter cited as *USP III*. All these men were closely connected with Philadelphia pharmacy; see England, *First Century*, 395ff.

54. [Procter], "Editorial Department," *AJP* 23 (1851): 297.

55. Ibid., *AJP* 28 (1856): 191. This second edition of the third revision allowed the committee to correct some of the mistakes in the first edition.

56. James Harvey Young, *Pure Food: Securing the Federal Food and Drugs Act of 1906* (Princeton, N.J.: Princeton Univ. Press, 1989), 6–10.

57. England, *First Century*, 133; Sonnedecker, *Kremers and Urdang's History of Pharmacy*, 198–200; see also chap. 7.

58. M. J. Bailey, "Communication [of the New York Examiner of Drugs]," *Proc APhA* 6 (1857): 101.

59. [Procter], "Editorial Department," *AJP* 27 (1855): 191.

60. Ibid., *AJP* 25 (1853): 187; Procter, George D. Coggeshall, and Edward S. Wayne, "Report of the Committee on Standards for Drugs," *Proc APhA* 4 (1855): 15; Procter, George D. Coggeshall, E. S. Wayne, "[Report of the Committee on Standards of Quality for Drug Examiners]," *Proc APhA* 5 (1856): 9, 25–26; see also Young, *Pure Food*, 10–17.

61. [Procter], "On the Means for Determining the Purity of Certain Chemicals and Drugs, and for Detecting Adulterations," *AJP* 23 (1851): 3.

62. Procter, "Remarks on Bogota and Other Cinchona Barks from New Grenada, in Their Relation to the Manufacture of Sulphate of Quinia," *AJP* 25 (1853): 306ff.

63. [Procter], "On Benzoic Acid from Urine," *AJP* 27 (1855): 23–24.

64. "Minutes of the Philadelphia College of Pharmacy," *AJP* 31 (1859): 282, 582; [L. V. Newton], "The Pharmacopoeia and Its Revision," *Drugg Circ* 3 (1859): 133.

65. "Proceedings of the National Convention for Revising the Pharmacopoeia of the U. States," *AJP* 32 (1860): 370–72, hereafter cited as "Proceedings, U. States"; [Procter], "Editorial Department," *AJP* 32 (1860): 376; Sonnedecker, *Kremers and Urdang's History of Pharmacy*, 260–67. Starting in 1880 more pharmacists than physicians sat on the revision committee (M. Clayton Thrush, "The United States Pharmacopoeia and Its Predecessors," *Drugg Circ* 51 [1907]: 49).

66. Franklin Bache, "On the Advantages That Would Accrue to English and American Pharmacy by the Adoption of a Single Uniform Pharmacopoeia for the British Empire," *AJP* 27 (1855): 11f.

67. Edward R. Squibb, "Weights and Measures of the Pharmacopoeia," *AJP* 32 (1860): 27–28; [Procter], "Editorial Department," *AJP* 32 (1860): 187, and *AJP* 33 (1861): 90.
68. "Proceedings, U. States," 374f.
69. [Procter], "Editorial Department," *AJP* 32 (1860): 376.
70. "Minutes of the Committee of Revision of the United States Pharmacopoeia, 1863," in dean's office, Philadelphia College of Pharmacy and Science, March 22, 1862; hereafter cited as "Minutes, 1863." This is the minutes book of the actual pharmacopoeial revision committee, not that of the College of Pharmacy.
71. "Minutes, 1863," where Bache reads Wood's letter to the committee, September 19, 1860, n.p.; see also August 16, 1862, n.p.
72. "Minutes, 1863," et seq.; [Procter], "Editorial Department," *AJP* 33 (1861): 384.
73. For example, see "Minutes, 1863," April 24, 1861. If he could not get down to Philadelphia, Squibb usually sent reports via mail; ibid., January 9, 1861. For an example of his expertise, see ibid., December 19, 1860.
74. [Procter], "Editorial Department," *AJP* 32 (1860): 571; "Minutes, 1863," March 29, 1862; see also May 10, 1862.
75. "Minutes, 1863," November 21, 1860.
76. Ibid., October 4, 1862.
77. Ibid., October 18, 1862. The essay appeared in the *USP* as part of its "Preliminary Notices."
78. Ibid., June 26, July 10, July 17, July 24, July 31, 1861.
79. Ibid., July 31, August 14, and August 21, 1861; May 17, May 24, May 31, July 5, July 12, 1862.
80. Ibid., January 23, April 10, 1861; see also May 3, 1862.
81. Edward Parrish, "On the Publication of the Revised Pharmacopoeia," *AJP* 34 (1862): 512–14; see also J. C., "The Pharmacopoeia of the United States of America. Fourth Decennial Revision . . . [review]," *American Journal of the Medical Sciences* 46 (1863): 433.
82. [Procter], "Editorial Department," *AJP* 35 (1863): 376.
83. "Minutes, 1863," January 24, 1863.
84. The monograph for dilute acetic acid was unchanged; Alfred B. Taylor, "Review of the United States Pharmacopoeia," *AJP* 35 (1863): 402n.
85. Procter, "Remarks on Some Preparations of the U.S. Pharmacopoeia," *AJP* 36 (1864): 298; [Procter], "Editorial Department," *AJP* 35 (1863): 282, 476; Taylor, "Review," 412, 420.
86. [Procter], "Editorial Department," *AJP* 36 (1864): 286.
87. Procter, "Lecture, Introductory to the Course on Practical Pharmacy," 245.
88. J. C., "Pharmacopoeia," 438.
89. Sargent, "On Fluid Extracts," 338.
90. Ibid., 341; [Procter], "Editorial Department," *AJP* 41 (1869): 93; Sonnedecker, *Kremers and Urdang's History of Pharmacy*, 328–29; Procter, "The Pharmacopoeia of 1870: Shall Its Authority Be More Generally Respected?" *AJP* 41 (1869): 289f.
91. Procter, "Pharmacopoeia of 1870," 290f.
92. Sonnedecker, *Kremers and Urdang's History of Pharmacy*, 267.

93. Oscar Oldberg, "Calomel," *AJP* 43 (1871): 154.
94. H. H. Rusby, "Fifty Years of Materia Medica," *Drugg Circ* 51 (1907): 40; Oldberg, "Calomel," 154; Sonnedecker, *Kremers and Urdang's History of Pharmacy*, 265–67.
95. [Procter], "Editorial Department," *AJP* 23 (1851): 389, emphasis added. Cf. [L. V. Newton], "Death of Dr. Franklin Bache," *Drugg Circ* 8 (1864): 67; and "Pharmacopoeia and Its Revision," 133.
96. For example, see Procter, "Test for the Presence of Alcohol in Chloroform," *AJP* 28 (1856): 213. Wood did not have Bache's chemistry skills.
97. Sonnedecker, *Kremers and Urdang's History of Pharmacy*, 279; Wood and Bache, *Dispensatory* (1865 ed.), 909.
98. Wood and Bache, *Dispensatory* (1865 ed.), 909–47. See also [Procter], "A Little Side-Talk with the 'Freshmen' in Pharmacy," *AJP* 43 (1871): 10.
99. *USP III*, 27.
100. Wood and Bache, *Dispensatory* (1865 ed.), 300–305.
101. Procter, "Pharmacopoeia of 1870," 289f.
102. Ibid.
103. George Urdang, "The Rescue of the U.S. Pharmacopoeia by Organized American Pharmacy in the Eighteen Seventies," *Am J Pharm Ed* 15 (1950): 172–84.
104. Sonnedecker, *Kremer and Urdang's History of Pharmacy*, 267.
105. Ibid., 329.

5. Editor of the *American Journal of Pharmacy*

1. Frederick Hoffmann, "A Century of American Pharmaceutical Literature and Journalism," *American Druggist and Pharmaceutical Record* 36 (1900): 164.
2. Daniels, *American Science*, 234; Frank Luther Mott, *A History of American Magazines, 1850–1865* (Cambridge, Mass.: Harvard Univ. Press, 1938), 92.
3. "Introduction," *Journal of the Philadelphia College of Pharmacy* 1 (1825): 3.
4. Daniel B. Smith, Benjamin Ellis, Robert E. Griffith, and Joseph Carson preceded Procter as editor.
5. Galen, "Philadelphia Correspondence of the Druggists' Circular," 137.
6. Higby, "Professionalism and the Nineteenth-Century American Pharmacist," 115–24.
7. England, *First Century*, 400–401; see chap. 4.
8. "Minutes of the Philadelphia College of Pharmacy," *AJP* 13 (1841): 80.
9. "Publishing Committee," *AJP* 13 (1841): i.
10. [Procter], "Elements of Chemistry, Including the Applications of the Science in the Arts by Thomas Graham [review]," *AJP* 15 (1843): 193–94.
11. Andrew Robertson, T. Smith, and H. Smith, "On Extract of Indian Hemp," *AJP* 19 (1847): 196–97.
12. J[oseph] C[arson], "Editorial Department," *AJP* 20 (1848): 253.
13. England, *First Century*, 400–401.
14. Carson wrote in a choppier style than Procter and liked to use the colon; Procter wrote longer sentences and relied more on the semicolon. A telltale sign of a Procter piece is "whilst," which he utilized in approximately half of his writings of a page or more.

15. C[arson], "Editorial Department," 253–56.

16. England, *First Century*, 401.

17. James Cassedy, "The Flourishing and Character of Early American Medical Journalism, 1797–1860," *Journal of the History of Medicine and Allied Health Sciences* 38 (1983): 148; "Minutes of the Philadelphia College of Pharmacy," *AJP* 43 (1871): 40–41; see also chap. 1.

18. Cassedy, "Early American Medical Journalism," 144–46.

19. [Procter, and Joseph Carson], "Editorial Department," *AJP* 22 (1850): 277.

20. [Procter], "Varieties," *AJP* 26 (1854): 272–79, 365–74, 460–70, 554–64; *AJP* 27 (1855): 74–82, 176–81, 273–78, 467–71, 558–65; *AJP* 28 (1856): 78–86, 175–87, 267–73, 365–74.

21. Ibid., *AJP* 30 (1858): 87; *AJP* 33 (1861): 379; *AJP* 34 (1862): 185; *AJP* 35 (1863): 470; *AJP* 40 (1868): 82; *AJP* 38 (1866): 182; see also chap. 3.

22. [Procter], "Gleanings," *AJP* 26 (1854): 9–17.

23. England, *First Century*, 278–79.

24. "Minutes of the Philadelphia College of Pharmacy," *AJP* 23 (1851): 284; [Procter], "Editorial Department," *AJP* 23 (1851): 91; and *AJP* 25 (1853): 575. Procter reduced the size of the index by combining the author and title lists.

25. [Procter], "Editorial Department," *AJP* 25 (1853): 473; "Minutes of the Philadelphia College of Pharmacy," *AJP* 26 (1854): 284. Although the figures are vague for this period, this seems to be when the journal's subscription list started to expand; "Minutes of the Philadelphia College of Pharmacy," *AJP* 41 (1869): 276.

26. [Procter], "Editorial Department," *AJP* 25 (1853): 473. For example, see H. P. Reynolds, "On Campbell's Method of Percolation," *AJP* 41 (1869): 525–27.

27. J[oseph] C[arson], "Editorial Department," *AJP* 20 (1848): 253. There is no mention in the College minutes of any serious conflicts between the publishing committee and Carson.

28. "Minutes of the Philadelphia College of Pharmacy," *AJP* 41 (1869): 274f.

29. At that time the *Journal* had about 500–600 regular paying subscribers (ibid., 274f.).

30. Frank Luther Mott, *A History of American Magazines, 1865–1885* (Cambridge, Mass.: Harvard Univ. Press, 1938), 4ff.

31. By 1861, 1,200 copies of the *Journal* were printed ("Minutes of the Philadelphia College of Pharmacy," *AJP* 41 [1869]: 276).

32. [Procter], "Editorial Department," *AJP* 33 (1861): 384.

33. Ibid., 478.

34. Parrish, "What Shall the Boys Do," 155; [L. V. Newton], "Dull Times," *Drugg Circ* 5 (1861): 85.

35. "Abstract of the Minutes of the Philadelphia College of Pharmacy," *AJP* 34 (1862): 275. In previous decades the *Journal* had produced a profit for the College.

36. Mott, *American Magazines, 1865–1885*, 6f.

37. "Abstract of the Minutes of the Philadelphia College of Pharmacy," *AJP* 35 (1863): 280f.

38. "Minutes of the Philadelphia College of Pharmacy," *AJP* 36 (1864): 273f.;

J. E. Howard, "Microscopical Researches on the Alkaloids, as Existing in Cinchona Bark," *AJP* 37 (1865): 346; "Minutes of the Philadelphia College of Pharmacy," *AJP* 37 (1865): 231–32.

39. "Minutes of the Philadelphia College of Pharmacy," *AJP* 38 (1866): 277; *AJP* 39 (1867): 278; [Procter], "Editorial Department," *AJP* 39 (1867): 87.

40. Robert Hare, "Criticisms and Suggestions Respecting Nomenclature," *AJP* 9 (1837): 1–16.

41. In 1861, for instance, the ratio was 36.5 percent original to 63.5 percent copied.

42. These calculations are based on examination of the content of pages in volumes of *AJP* for 1847, 1851, 1861, and 1870. Original material included articles, notes, letters to the editor, reviews, inaugural essays, College reports, and editorial departments; reprinted material included articles and abstract departments (Miscellany, Varieties, and Gleanings).

43. "On Chemical Substitutes for the Fermentation of Bread," *AJP* 26 (1854): 42–46; W. Munsel, "On the Preparation of Dammara Varnish," *AJP* 26 (1854): 317–19; David Stuart [Stewart], "On Nascent Manures," *AJP* 27 (1855): 246–51; William Hooker and Daniel Hanbury, "Botanical and Pharmacological Inquiries and Desiderata," *AJP* 32 (1860): 49–58; L[ouis] Pasteur, "On the Origin of Ferments," *AJP* 33 (1861): 165–70; L[ouis] Pasteur, "Researches on the Molecular Dissymmetry of Natural Organic Products," *AJP* 34 (1862): 1–16, 97–112; L[ouis] Pasteur, "Researches on Acetic Fermentation," *AJP* 37 (1865): 343–44; W. H. Perkins [*sic*], "On Mauve, or Aniline Purple," *AJP* 36 (1864): 171–73; "Relation of the Chemical Constitution and Physiological Action of Medicine. Addition of Iodide of Methyl to Vegetable Alkaloids," *AJP* 40 (1868): 440–42; Geo. J. Engelmann, "Chloral—A New Anaesthetic," *AJP* 41 (1869): 447–48.

44. [Procter], "Editorial Department," *AJP* 32 (1860): 89.

45. Ibid., *AJP* 33 (1861): 89.

46. For example, R. H. Stabler, "Practical Observations," *AJP* 23 (1851): 121.

47. Cassedy, "Flourishing," 144. Procter's editorial protégé, John Maisch, followed his example when he became the journal's editor in 1870.

48. John T. Plummer, "Proximative Analysis of a Concretion of Hairs Found in the Esophagus of a Slaughtered Ox," *AJP* 25 (1853): 101–4.

49. England, *First Century,* 252.

50. When Procter became editor in 1850 the circulation was about 350; when he retired in 1871 it was around 1,200 ("Minutes of the Philadelphia College of Pharmacy," *AJP* 41 [1869]: 275–76).

51. [Procter], "Citrate of Iron and Strychnia," *AJP* 31 (1859): 23–24.

52. [Procter], "Editorial Department," *AJP* 25 (1853): 187; S. P. Peck, "Letter Relating to the Sale of Poisons in Vermont," *AJP* 25 (1853): 207; [Procter], "Citrate," 23–24.

53. Ironically, one of the pharmacists who contributed to the *Journal* disliked its supposed lack of practical information (Wm. L. Turner, "Note on Lozenge Cutting," *AJP* 39 [1867]: 206; "The Doctor and the Apothecary," *AJP* 43 [1871]: 149–51; "The Need of Practical Information in Our Pharmaceutical Publications," *AJP* 44 [1872]: 533).

54. [Procter], "Editorial Department," *AJP* 39 (1867): 88.

55. Ibid., 88–89.
56. Joseph Harrop, "On Syrup of Citric Acid of the Pharmacopoeia," *AJP* 41 (1869): 393; H. P. Reynolds, "On Mistura Cretae," *AJP* 42 (1870): 391–92; Geo. W. Kennedy, "Tincture of Nux Vomica," *AJP* 42 (1870): 392–93; W. Ranstead Jones, "Chalk Mixture," *AJP* 42 (1870): 393.
57. [Procter], "Editorial Department," *AJP* 39 (1867): 88. According to Procter, this statement appeared on the cover of the journal.
58. D. L. Phares, "Virburnum Prunifolium in the Treatment of Threatened Abortion," *AJP* 39 (1867): 260n.
59. Hoffmann, "Pharmaceutical Literature," 163–64. L. V. Newton, a New York physician, bought the *Druggists' Circular* in 1858, one year after its founding. Because of his fragile health, he decided to go into journalism. He edited the *Circular* until 1880.
60. *Drugg Circ* 10, no. 4 (1866).
61. [L. V. Newton], "College of Pharmacy," *Drugg Circ* 2 (1858): 197; "Saluta-tory," *Drugg Circ* 3 (1859): 13. He wrote in a sweeping, conversational style, usually with a hortatory message; see [L. V. Newton], "Sunday Store-keeping," *Drugg Circ* 4 (1860): 129–30; "Concerning the Art of Putting Things," *Drugg Circ* 5 (1861): 37–38; "Looking through a Brick," *Drugg Circ* 8 (1864): 87–88.
62. Lucy M. Salmon, *The Newspaper and the Historian* (New York: Oxford Univ. Press, 1923), 249ff.
63. [Procter and Joseph Carson], "Editorial Department," *AJP* 20 (1848): 349–52.
64. Ibid., *AJP* 21 (1849): 96.
65. Ibid., *AJP* 22 (1850): 279–87.
66. Ibid., *AJP* 21 (1849): 191, 382–84.
67. [Procter], "Editorial Department," *AJP* 23 (1851): 186f., 286–92.
68. See chap. 6; [Procter], "Editorial Department," *AJP* 24 (1852): 85, 185–86, 281–84, 381–85; *AJP* 25 (1853): 369–71; *AJP* 26 (1854): 286–87.
69. Ibid., *AJP* 25 (1853): 87, 369–70; *AJP* 26 (1854): 90, 169–70.
70. Ibid., *AJP* 26 (1854): 567; *AJP* 27 (1855): 185–86; *AJP* 29 (1857): 479.
71. Ibid., *AJP* 28 (1856): 91.
72. Berman, "A Striving," 17.
73. For example, the St. Louis Pharmaceutical Association and the Richmond Pharmaceutical Society.
74. [Procter], "Editorial Department," *AJP* 30 (1858): 186–91, 577–80; *AJP* 31 (1859): 88–94, 285–88, 384–92.
75. See chaps. 2 and 4; [Procter], "Editorial Department," *AJP* 35 (1863): 187–89; *AJP* 37 (1865): 315; *AJP* 38 (1866): 92, 282–85; *AJP* 40 (1868): 284.
76. [Procter], "Editorial Department," *AJP* 34 (1862): 379; *AJP* 35 (1863): 569–71; *AJP* 36 (1864): 93, 461–62; *AJP* 37 (1865): 74, 318.
77. Ibid., *AJP* 38 (1866): 189, 278–80, 281, 381, 476–77, 480, 560–74; *AJP* 39 (1867): 187, 565–66; *AJP* 41 (1869): 371–73, 379.
78. Ibid., *AJP* 41 (1869): 186.
79. Ibid., 377.
80. Ibid., *AJP* 42 (1870): 82, 378; *AJP* 43 (1871): 94, 141.
81. Ibid., *AJP* 42 (1870): 85.

82. Ibid., 82.
83. "Minutes of the Philadelphia College of Pharmacy," *AJP* 41 (1869): 274f. See also [Procter], "Editorial Department," *AJP* 41 (1869): 87; *AJP* 43 (1871): 42, 94.
84. "Minutes of the Philadelphia College of Pharmacy," *AJP* 41 (1869): 274.
85. [Procter], "Editorial Department," *AJP* 41 (1869): 380.
86. "Minutes of the Philadelphia College of Pharmacy," *AJP* 33 (1861): 284. This comment is from a publishing committee report that Procter probably wrote.
87. Urdang, "Edward Parrish," 225.
88. "Minutes of the Philadelphia College of Pharmacy," *AJP* 45 (1873): 225–32. Procter's name is not given as author of this published obituary, but he called it "my memoir" in a letter he wrote to Albert Ebert (A2: Procter, KRF).
89. This may have been modeled on the "Supplement" of the *Druggists' Circular;* [Procter], "Editorial Department," *AJP* 42 (1870): 381.
90. Ibid., 572. Procter suggested the new position.
91. "Minutes of the Philadelphia College of Pharmacy," *AJP* 43 (1871): 40.
92. [Procter], "Editorial Department," *AJP* 43 (1871): 94.
93. Ibid., 185–92.
94. "Minutes of the Philadelphia College of Pharmacy," *AJP* 43 (1871): 40–41.
95. [John M. Maisch], "Ferrated Elixir of Cinchona," *AJP* 43 (1871): 219.
96. [John M. Maisch], "Editorial Department," *AJP* 43 (1871): 233.

6. Pioneering Pharmaceutical Educator

1. Sonnedecker, "Pharmaceutical Education," 435, 471. David Stewart of Baltimore (1813–99) preceded Procter as the first holder of a free-standing chair in pharmacy in the United States. Stewart, a physician, taught just a handful of pharmacists during his brief tenure (1844–46), however, and did not contribute significantly to pharmaceutical education (George E. Osborne, "David Stewart, M.D.: First American Professor of Pharmacy," *Am J Pharm Ed* 23 [1959]: 219–30). Procter, always conscious about priority, gave Stewart credit as the first (A. P. Sharp in Remington, "Procter Reminiscences," 556).
2. Procter, "Valedictory Address," 204–5.
3. Sonnedecker, "Pharmaceutical Education," 28.
4. Ibid., 275.
5. Ibid., 272; [Procter, E. Parrish, D. Stewart, and J. Meakim], "[Report of the Committee on Education] to the Pharmaceutists of the United States," *Proc APhA* 3 (1854): 14–15; Parrish, "American Pharmacy," 117.
6. [Procter, Parrish, Stewart, and Meakim], "To the Pharmaceutists," 14–15.
7. Benjamin, *Philadelphia Quakers,* 7, 51.
8. See chap. 1; [Procter], "A Little Side-Talk with the 'Freshmen,'" 8.
9. Sonnedecker, *Kremers and Urdang's History of Pharmacy,* 189f., 232.
10. Wood left the College faculty in 1836 to join the Medical Department of the University of Pennsylvania. Carson succeeded him in the chair of materia medica and pharmacy.
11. Procter, "Lecture, Introductory to the Course on Practical Pharmacy," 244–45.

12. Sonnedecker, "Pharmaceutical Education," 423.
13. Ibid., 461; Urdang, "The Part of Doctors of Medicine in Pharmaceutical Education," 549.
14. In his introduction to the course on materia medica and pharmacy, which Procter probably heard, Wood virtually ignored pharmacy; George B. Wood, "Lecture Introductory to the Course on Materia Medica and Pharmacy," *AJP* 8 (1836): 286–300.
15. Sonnedecker, "Pharmaceutical Education," 433–34.
16. "Minutes of the College of Pharmacy," *AJP* 18 (1846): 147.
17. Ibid., 150.
18. Ibid., 154.
19. Ibid., 147.
20. Ibid., 155; [Bullock], "Memoir of William Procter," 522.
21. Procter, "Lecture, Introductory to the Course on Practical Pharmacy," 253f.
22. Ibid., 245–46.
23. Ibid., 254; "Commencement of the Philadelphia College of Pharmacy," *Drugg Circ* 5 (1861): 85.
24. C. Lewis Diehl, "Notes of Lectures Taken at the Philada. College of Pharmacy," ms. student notebook, Archives, American Pharmaceutical Association, Washington, D.C.; [Bullock], "Memoir of William Procter," 523.
25. [Bullock], "Memoir of William Procter," 519; Bullock had access to Procter's diary (see chap. 1).
26. [Procter], "Editorial Department," *AJP* 36 (1864): 288.
27. England, *First Century,* 83; Hancock, "William Procter, Jr., the Father of American Pharmacy," 18; James T. Shinn, "Tribute to the Memory of Professor Procter," *AJP* 77 (1905): 40.
28. Adolph W. Miller in Remington, "Tributes," 37.
29. Remington, "Procter Reminiscences," 553.
30. Edward Parrish, "A Critical Notice of the Proceedings of the Pharmaceutical Association," *Drugg Circ* 2 (1858): 232; Procter, "Syllabus of a Course of Study," *Proc APhA* 7 (1858): 134.
31. Procter, "Can Practical Pharmacy be Taught Effectively by Lectures," *AJP* 43 (1871): 99; Sonnedecker, "Pharmaceutical Education," 448 n. 181.
32. [Procter], "Editorial Department," *AJP* 23 (1851): 192.
33. Remington, "Procter Reminiscences," 553; student notebook by Diehl shows drawings of demonstrations, for example, pp. 201, 230, and 241; "Minutes of the Philadelphia College of Pharmacy," *AJP* 45 (1873): 230; see also letter from Procter to Albert Ebert dated April 20, 1873, in A2: Procter, KRE
34. This advice comes from Procter's "Syllabus," which directly paralleled his course (Procter, "Syllabus," 111, 134; compare Diehl's student notebook).
35. Procter, "Syllabus," 135, 137; see also Diehl's student notebook, 255f.
36. Diehl's student notebook, 20f., 36f.
37. Ibid.
38. Remington, "Procter Reminiscences," 554f.; see chap. 1.
39. [Procter], "Editorial Department," *AJP* 33 (1861): 576.
40. "Minutes of the Philadelphia College of Pharmacy," *AJP* 45 (1873): 230.
41. [Bullock], "Memoir of William Procter," 528–29.

42. "Death of Professor William Proctor [*sic*], Jun.," *The Chemist and Druggist* 15 (1874): 77.

43. The death certificate states that he died early on the morning of the tenth; he gave his last lecture on February 9. Remington, "Procter Reminiscences," 555; [Bullock], "Memoir of William Procter," 529; Frederick B. Power, student notebook, Kremers Reference Files, University of Wisconsin-Madison School of Pharmacy, Madison, Wisc.

44. There were no laws that required pharmacists to have a diploma to practice (Sonnedecker, "Pharmaceutical Education," 282–96).

45. Procter, Charles Ellis, and A. P. Sharp, "[Report of the Committee on Textbooks]," *Proc APhA* 5 (1856): 16.

46. "Minutes of the Philadelphia College of Pharmacy," *AJP* 18 (1846): 154.

47. Sonnedecker, *Kremers and Urdang's History of Pharmacy,* 282f. For the situation in medicinal chemistry, see James Whorton, "Chemistry," in *The Education of American Physicians,* ed. Ronald L. Numbers (Berkeley: Univ. of California Press, 1980), 80.

48. George Urdang, "The Development of the Pharmaceutical Text," *Am J Pharm Ed* 8 (1944): 331f.; Armin Wankmueller, "Deutschsprachige Lehrbuecher der pharmazeutischen Technologie von 1845 bis 1945," *Deutsche Apotheker Zeitung* 122 (1982): 1236.

49. John Mark Scott, "K. F. Mohr, Father of Volumetric Analysis," *Chymia* 3 (1950): 195–96.

50. Ibid., 191–92, 197.

51. Mohr and Redwood, *Practical Pharmacy,* v.

52. Ibid., vi.

53. Ibid., 223; Karl F. Mohr, *Lehrbuch der pharmaceutischen Technik* (Braunschweig: Friedrich Bieweg, 1847), 237–41.

54. Mohr and Redwood, *Practical Pharmacy,* 253–56.

55. Ibid., 327–70.

56. Procter agreed to edit the book for $250 and six copies; "Lea and Blanchard Cost Book, 1838–1853," Lea and Febiger Collection, Historical Society of Pennsylvania, Manuscripts Department, p. 141.

57. Mohr, Redwood, and Procter, *Practical Pharmacy,* iii.

58. William F. Norwood, *Medical Education in the United States before the Civil War* (Philadelphia: Univ. of Pennsylvania Press, 1944), 398; Hellmut Lehmann-Haupt, *The Book in America,* 2d ed. (New York: Bowker, 1951), 108–13.

59. *Catalogue of Blanchard and Lea's Medical and Surgical Publications* (Philadelphia: Blanchard and Lea, 1853), 6, 15, 24.

60. Mohr and Redwood, *Practical Pharmacy,* 46–134.

61. Mohr, Redwood, and Procter, *Practical Pharmacy,* 94–99, iii–v, 368–85; Sonnedecker, "Pharmaceutical Education," 517.

62. Mohr, Redwood, and Procter, *Practical Pharmacy,* 50–55.

63. Ibid., 413, 421.

64. Ibid., 414–16, 419–22.

65. Ibid., 504.

66. Ibid., 506.

67. Ibid., 542.

68. Ibid., 543.

69. Procter, *Valedictory Address to the Graduates of the Philadelphia College of Pharmacy . . . 1852* (Philadelphia: Merrihew and Thompson, 1852), 8–9.

70. Parrish's realistic approach is typified by the appendix to the *Introduction* that describes the preparation of "physician's outfits," and some recipes for "the most Important Patent Medicines" (Parrish, *Introduction to Practical Pharmacy*, 509–14, 522–25; see also, Urdang, "Edward Parrish, A Forgotten Pharmaceutical Reformer," 223–32).

71. Procter, Ellis, and Sharp, "[Report]," 16.

72. "[Minutes of the Annual Meeting]," *Proc APhA* 6 (1857): 13.

73. Procter, "Syllabus," 104.

74. [L. V. Newton,] "Good News for the Rising Generation of Pharmaceutists," *Drugg Circ* 2 (1858): 217.

75. Parrish, "Critical Notice," 232.

76. Sonnedecker, "Pharmaceutical Education," 435f.; see also "Notes on Pharmaceutical Education," *Proc APhA* 40 (1892): 320–22.

77. A comparison of the syllabus and student notebooks from the 1860s and 1870s shows that Procter kept the same arrangement of his subject, only changing the particulars as new discoveries came to light (Diehl, student notebook, and Power, student notebook).

78. In contrast, the description of "a true drugstore" by L. V. Newton of the *Druggists' Circular* omitted on-site manufacture, which Procter emphasized; [Newton], "What is a Drug Store Proper?" 278.

79. Procter, "Thoughts on 'Manufacturing Pharmacy,'" 516.

80. [Procter], "Editorial Department," *AJP* 38 (1866): 187–88.

81. Ibid., *AJP* 40 (1868): 469–70; *AJP* 41 (1869): 83–85.

82. Ibid., *AJP* 41 (1869): 285.

83. "Philadelphia College of Pharmacy," *AJP* 43 (1871): 175f.

84. Parrish designed his school mainly for physicians, but some pharmacy apprentices, especially from wholesale druggist firms, took his practical course (Sonnedecker, "Pharmaceutical Education," 362 n. 87, 441).

85. [Procter], "Observations on the Present State of Pharmacy in England," *AJP* 22 (1850): 15.

86. [Procter], "Editorial Department," *AJP* 25 (1853): 373; *AJP* 31 (1859): 584.

87. Ibid., *AJP* 38 (1866): 187.

88. Ibid., 188; compare Sonnedecker, "Pharmaceutical Education," 361–66.

89. Procter to Ebert, February 28, 1869, A2: Procter, KRF. Ebert, a student of Procter at the College during the early 1860s, went to Germany in the mid-1860s and earned a Ph.D. in pharmaceutical chemistry under Georg C. Wittstein.

90. England, *First Century*, 162; [Procter], "Editorial Department," *AJP* 42 (1870): 474; *AJP* 43 (1871): 186; "Pharmaceutical Colleges and Associations," *AJP* 44 (1872): 522f.

91. [Procter], "Editorial Department," *AJP* 39 (1867): 94.

92. Procter to Ebert, February 28, 1869, A2: Procter, KRF

93. [Procter], "Editorial Department," *AJP* 41 (1869): 472.

94. Ibid., 575.

95. Ibid.

96. Sonnedecker, "Pharmaceutical Education," 205–6, 521. See also Glenn A.

Sonnedecker, "The Conference of Schools of Pharmacy," *Am J Pharm Ed* 18 (1954): 389–95.

97. "Minutes of the Convention of Delegates from Colleges and Societies of Pharmacy . . . Relative to Pharmaceutical Education," *AJP* 42 (1870): 500ff.

98. Ibid., 502–4.

99. [Procter], "Editorial Department," *AJP* 43 (1871): 44. The trustees of the Philadelphia College of Pharmacy adopted a similar position, deciding that a student without store experience would only receive a certificate of proficiency in pharmacy ("Pharmaceutical Colleges and Associations," *AJP* 45 [1873]: 378).

100. Procter, "Can Practical Pharmacy," 98.

101. Ibid., 99–101. This clash between society and university pharmacy schools was similar to those in medical education earlier in the century (William Rothstein, *American Physicians in the Nineteenth Century* [Johns Hopkins Univ. Press, 1977], 101–21).

102. "Minutes of the Nineteenth Annual Meeting," *Proc APhA* 19 (1871): 29.

103. Ibid., 30–33.

104. Ibid., 33–34.

105. Ibid., 34. This committee of ten included five men connected with the Philadelphia College of Pharmacy: Procter, Parrish, Diehl, John F Hancock, and Ebert.

106. Ibid., 47.

107. Ibid., 48.

108. A. B. Prescott, "Pharmaceutical Education," *Proc APhA* 19 (1871): 425–29. As Prescott pointed out, there were no laws at that time regulating the training of apprentices in pharmacy or medicine.

109. "Minutes," *Proc APhA* 19 (1871): 96.

110. It would not be until 1883, over ten years later, that a state university would administer a pharmacy course acceptable to the old-line schools. At the University of Wisconsin, students could attend classes without practical experience, but needed four years of apprenticeship for a diploma (Sonnedecker, *Kremers and Urdang's History of Pharmacy,* 235).

111. For example, see [Procter], "Editorial Department," *AJP* 26 (1854): 474–75; *AJP* 37 (1865): 400; *AJP* 42 (1870): 475.

112. Ibid., *AJP* 38 (1866): 479–80; Colcord, "Pharmaceutical Colleges and Associations," 136; Ebert correspondence, A2: Procter, KRF

113. Parrish recognized the trend, but failed to suggest any alternative to the apprenticeship-finishing school model (Parrish, "On the Relations of the Several Classes of Druggists and Pharmacists to the Colleges of Pharmacy," 532; cf. Melvin R. Gibson, "Scope [of Pharmacy]," in *Remington's Pharmaceutical Sciences* [Easton, Penn.: Mack Publishing, 1970], 3–5).

114. Sonnedecker, "Pharmaceutical Education," 135.

115. Ibid., 461.

116. Parrish, "American Pharmacy," 216. Parrish's rather negative views on the medical profession are especially interesting inasmuch as his father was Joseph Parrish, one of Philadelphia's most respected physicians.

117. Kremers Reference Files C31(a)I: [University of] Wisconsin [School of Pharmacy].

118. England, *First Century*, 205–8.
119. Ibid., 407–8.
120. Other Procter students who later taught pharmacy included Lucius E. Sayre, Albert Ebert, Emlen Painter, William T. Wenzell, and C. Lewis Diehl.

7. Procter and Pharmaceutical Organizations

1. "Minutes of the Twenty-second Annual Meeting," *Proc APhA* 22 (1874): 489.
2. Ibid., 538.
3. Urdang, "The Part of Doctors of Medicine in Pharmaceutical Education," 547.
4. Ebert, "The Father of American Pharmacy," 155.
5. [Procter], "Editorial Department," *AJP* 38 (1866): 480.
6. Ibid., *AJP* 28 (1856): 188.
7. Ibid., *AJP* 30 (1858): 185; *AJP* 32 (1860): 90.
8. For example, see ibid., *AJP* 23 (1851): 91–92; *AJP* 26 (1854): 381, 473–77; *AJP* 28 (1856): 89–90; *AJP* 42 (1870): 188.
9. Ibid., *AJP* 26 (1854): 92; *AJP* 31 (1859): 287; and Procter to Ebert, September 20, 1868, A2: Procter, KRF.
10. "Minute Book of the Trustees of the Philada. College of Apothecaries [Pharmacy]," [Vol. 1, 1821–40], ms. bound volume, in rare book collection, Philadelphia College of Pharmacy and Science, 373. For more on Duhamel, see chap. 1.
11. Ibid., 384 et seq.; [Bullock], "Memoir of William Procter," 530.
12. "Abstract of the Minutes of the Philadelphia College of Pharmacy," *AJP* 34 (1862): 277–78.
13. Pharmaceutical College and Associations: Philadelphia College of Pharmacy," *AJP* 44 (1872): 39; see also letters from Procter to Ebert in A2: Procter, KRF. For a description of the troubles of the Chicago College of Pharmacy, see M. M. Weinstein, M. B. Mrtek, R. L. Lambert, and R. G. Mrtek, ". . . From These Ashes," *Pharmacy in History* 15 (1973): 54–65, 107–16.
14. "Charter and Laws of the Philadelphia College of Pharmacy," *AJP* 28 (1856): 282.
15. "Minutes of the Philadelphia College of Pharmacy," *AJP* 39 (1867): 279.
16. England, *First Century*, 342; [Bullock], "Memoir of William Procter," 530; "Minutes of the Philadelphia College of Pharmacy," *AJP* 44 (1872): 467.
17. Urdang implied that Procter purposefully stayed in these positions to keep Parrish out (Urdang, "Edward Parrish, A Forgotten Pharmaceutical Reformer," 223–32).
18. England, *First Century*, 132; Curt Wimmer, *The College of Pharmacy of the City of New York* (New York, 1929), 192ff.
19. Sonnedecker, *Kremers and Urdang's History of Pharmacy*, 198–99; [Procter], "Editorial Department," *AJP* 25 (1853): 89–90. For a full description of events leading up to the passage of the law, see Young, *Pure Food*, 3–17.
20. [Procter], "Editorial Department," *AJP* 24 (1852): 86.
21. [Procter], "Observations on the Present State of Pharmacy in England," 18.

22. *Minutes of the Convention of Pharmaceutists and Druggists Held in the City of New York, October 15, 1851* (Philadelphia: Merrihew and Son, 1865), 4.
23. "Proceedings of the Convention of Pharmaceutists and Druggists, Held in the City of New York, Oct. 15, 1851," *AJP* 24 (1852): 24.
24. Ibid., 25–26.
25. George D. Coggeshall (1809–91) was one of the founders and leaders of the New York College of Pharmacy. Samuel Colcord ran a successful wholesale drugstore in Boston and was prominent in the Massachusetts College of Pharmacy.
26. [Procter], "Editorial Department," *AJP* 24 (1852): 186.
27. Ibid., 282.
28. Ibid., 283–84.
29. Procter, Samuel M. Colcord, and George D. Coggeshall, "[Report of the Executive Committee] to the Convention of Pharmaceutists Assembled at Philadelphia, October 6th, 1852," *Proc APhA* 1 (1852): 7–8.
30. Ibid., 11–12.
31. Robert W. McCluggage, *A History of the American Dental Association* (Chicago, 1959), 49–52.
32. "Minutes of the Philadelphia College of Pharmacy," *AJP* 25 (1853): 278; Procter, Colcord, and Coggeshall, "[Report of the Executive Committee]," 8.
33. "Proceedings of the American Pharmaceutical Convention," *AJP* 25 (1853): 483–84.
34. "Minutes of the Eighteenth Annual Meeting," *Proc APhA* 18 (1870): 83.
35. "[Minutes of the Fourth Annual Meeting,]" *Proc APhA* 4 (1855): 12, 40.
36. [Procter], "Editorial Department," *AJP* 26 (1854): 287.
37. "Minutes of the Fifteenth Annual Meeting," *Proc APhA* 15 (1867): 35.
38. "Minutes of the Eleventh Annual Meeting," *Proc APhA* 11 (1864): 43–47; "Minutes of the Fourteenth Annual Meeting," *Proc APhA* 14 (1866): 70–74; "Minutes of the Twentieth Annual Meeting," *Proc APhA* 20 (1872): 105–9.
39. Procter, "Prefatory Note," *Proc APhA* 9 (1860): xv; see also "Minutes of the Nineteenth Annual Meeting," *Proc APhA* 19 (1871): 93.
40. I. J. Grahame, "The Process of Percolation or Displacement," *Proc APhA* 7 (1858): 285–94.
41. Charles T. Carney, "Paraffin—Its Substitution for Wax in Cerates," *Proc APhA* 9 (1860): 163–65.
42. Procter, "Remarks on Dialysis," *Proc APhA* 12 (1864): 264–70.
43. "Proceedings of the American Pharmaceutical Convention," *AJP* 25 (1853): 481ff. Although the *Journal*'s readership during the 1850s was under one thousand, this was significantly larger than the hundred or so pharmacists and physicians who were members of the association.
44. [Procter], "Editorial Department," *AJP* 25 (1853): 87; *AJP* 26 (1854): 169; *AJP* 27 (1855): 379, 479; *AJP* 28 (1856): 381; *AJP* 41 (1869): 469, 569.
45. Ibid., *AJP* 33 (1861): 191.
46. Ibid., 287.
47. H. T. Kiersted, "American Pharmaceutical Association: Notice of Postponement," *AJP* 33 (1861): 377.
48. [Procter], "Editorial Department," *AJP* 33 (1861): 382.
49. Ibid., *AJP* 34 (1862): 93.

50. Ibid., 491.
51. Ibid., *AJP* 35 (1863): 373; see also ibid., *AJP* 36 (1864): 380, 461, 541.
52. Ibid., *AJP* 38 (1866): 186–92.
53. Procter's attitudes toward several such controversies have been discussed in previous chapters dealing with pharmaceutical practice, science, standards, journalism, and education.
54. "Minutes of the Annual Meeting," *Proc APhA* 9 (1860): 57–58, 63; "Minutes of the Sixteenth Annual Meeting," *Proc APhA* 16 (1868): 64.
55. Edward Parrish, "The Pharmaceutical Convention of 1852," *AJP* 24 (1852): 134.
56. Parrish, "American Pharmacy," 290.
57. "Notice: To the Pharmaceutists and Druggists of the United States," *AJP* 27 (1855): 283.
58. [Procter], "Editorial Department," *AJP* 36 (1864): 541.
59. "Minutes," *Proc APhA* 16 (1868): 92, 102–4.
60. "Minutes of the Seventeenth Annual Meeting," *Proc APhA* 17 (1869): 100f.
61. Ibid., 103.
62. Ibid., 105.
63. Ibid., 111–12.
64. [Procter], "Editorial Department," *AJP* 30 (1858): 281–82; *Proc APhA* 9 (1860): 49; see also chap. 2.
65. "Minutes," *Proc APhA* 14 (1866): 51; see also *Proc APhA* 17 (1869): 61–75.
66. "Minutes," *Proc APhA* 16 (1868): 75.
67. Ibid., 109–10; Sonnedecker, *Kremers and Urdang's History of Pharmacy,* 215–17.
68. "Minutes," *Proc APhA* 17 (1869): 61.
69. Ibid., 84–86.
70. "Draft of a Proposed Law to Regulate the Practice of Pharmacy and the Sale of Poisons," *AJP* 41 (1869): 562–69.
71. [Procter], "Editorial Department," *AJP* 43 (1871): 141–42.
72. Ibid., *AJP* 26 (1854): 286.
73. Ibid., *AJP* 30 (1858): 381; "Minutes," *Proc APhA* 7 (1858): 25.
74. [Procter], "Editorial Department," *AJP* 35 (1863): 282; Procter, "American Pharmaceutical Association," *AJP* 35 (1863): 372.
75. "Minutes," *Proc APhA* 14 (1866): 66; [Procter], "Editorial Department," *AJP* 40 (1868): 378f.
76. "Minutes," *Proc APhA* 17 (1869): 87–92.
77. "Minutes of the Philadelphia College of Pharmacy," *AJP* 42 (1870): 564f.; [Procter], "Editorial Department," *AJP* 42 (1870): 376. For similar reasons, the AMA met in New Orleans in 1870.
78. "Minutes," *Proc APhA* 19 (1871): 107–11; "Minutes," *Proc APhA* 20 (1872): 98–102; "Minutes of the Twenty-first Annual Meeting," *Proc APhA* 21 (1873): 26. By 1873 the association had grown to include about 1,200 members.
79. "Minutes of the Philadelphia College of Pharmacy," *AJP* 45 (1873): 507.
80. "Minutes," *Proc APhA* 9 (1860): 30.
81. [Procter], "Editorial Department," *AJP* 34 (1862): 565.
82. Ibid., *AJP* 38 (1866): 476.
83. Sonnedecker, *Kremers and Urdang's History of Pharmacy,* 198ff.
84. "Minutes," *Proc APhA* 16 (1868): 73.

85. "Minutes of the Nineteenth Annual Meeting," *Proc APhA* 19 (1871): 75.
86. See Procter's memoir of Parrish in "Minutes of the Philadelphia College of Pharmacy," *AJP* 45 (1873): 225f.; see also Edward Parrish, "A Plea for the Handmaiden," *Proc APhA* 11 (1863): 271–73.
87. "Minutes," *Proc APhA* 17 (1869): 98; "Minutes," *Proc APhA* 15 (1867): 34–122; "Minutes," *Proc APhA* 9 (1860): 42; "Minutes," *Proc APhA* 20 (1872): 80, 82–83.
88. Lawrence G. Blochman, *Doctor Squibb* (New York: Simon and Schuster, 1958), 152–54.
89. Compare [Procter], "Editorial Department," *AJP* 26 (1854): 286–88, with Parrish, "American Pharmacy," 291f.
90. [Procter], "Editorial Department," *AJP* 29 (1857): 379.

8. Concluding Remarks

1. [Procter], "Observations on the Present State of Pharmacy in England," 2.
2. England, *First Century*, 184f.; "Procter Memorial Unveiled," *Journal of the American Pharmaceutical Association* (Prac. Pharm. Ed.) 2 (1941): 167.
3. "Pharmaceutical Colleges and Associations: Chicago College of Pharmacy," *AJP* 46 (1874): 140.
4. "Seventy-first Annual Meeting of the American Pharmaceutical Association . . . Abstract of Minutes," *Journal of the American Pharmaceutical Association* 12 (1923): 893–94.
5. Gregory J. Higby, ed., *One Hundred Years of the National Formulary* (Madison, Wisc.: American Institute of the History of Pharmacy, 1989).
6. [Bullock], "Memoir of William Procter," 517.
7. [Procter], "Necrology [Augustine Duhamel]," 316.

Appendix A.

1. The speech was published in French in the *Compte Rendu des Congrès Pharmaceutique Réunis, en Août 1867*, . . . (Paris: Congrès International des Associations et Sociétés de Pharmaciens, 1868), 55–61; in A2: Procter, KRF

Appendix B.

1. [Procter], "Editorial Department," *AJP* 26 (1854): 171.
2. [Procter], "Gelseminum *versus* Jasminum," *AJP* 40 (1868): 30.
3. [Procter], "Editorial Department," *AJP* 36 (1864): 464.
4. Ibid., *AJP* 30 (1858): 187.
5. Ibid., *AJP* 31 (1859): 93–94.

Appendix C.

1. Procter, "Note on Adulterated Powdered Socotrine Aloes," *AJP* 25 (1853): 99–101.

Appendix D.

1. Procter, "On Fluid Extract of Rhubarb and Senna," *AJP* 25 (1853): 23–24.

Bibliography

Abbreviations

AJP   *American Journal of Pharmacy*
Am J Pharm *American Journal of Pharmaceutical Education*
Drugg Circ *American Druggists' Circular and Chemical Gazette*
KRF   Kremers Reference Files, F. B. Power Pharmaceutical Library,
            University of Wisconsin-Madison School of Pharmacy
Proc APhA   *Proceedings of the American Pharmaceutical Association*

Published Works of William Procter, Jr.

This listing includes all published works for which Procter was the principal author. Because a large proportion of the pieces were regular departments in the periodical he edited, the *American Journal of Pharmacy,* these are grouped together by title. Often Procter's name did not appear in the byline, but was replaced by "W. P., Jr." or "By the Editor." For the sake of simplicity these pseudonyms are omitted. Reprinted articles are not included, and the *American Journal of Pharmacy* is given priority over the *Proceedings of the American Pharmaceutical Association* when the same paper appeared in both.

"On Lobelia Inflata," *AJP* 9 (1837): 98–109.
"Some Remarks on the Oil of Wild Cherry Bark," *AJP* 9 (1837): 298–304.
"Observations on Some of the Camphoriferous Essential Oils, and on the Resins Evolved from Some of the Volatile Oils, by Their Reaction with Sulphuric Acid," *AJP* 10 (1838): 17–24.
"Observations on Amygdaline, Demonstrative of its Existence in Several Species of Natural Division of Plants Amygdaleae," *AJP* 10 (1838): 188–200.
"Observations on the Carbonate and Protomuriate of Iron," *AJP* 10 (1838): 272–75.

235

"Observations on Xerophyllum Setifolium," *AJP* 11 (1839): 183–86.

"Observations on the Method of Displacement," *AJP* 11 (1839): 189–201 [with Augustine Duhamel].

"Observations on Dextrine and Diastase," *AJP* 11 (1840): 265–79.

"Observations on Lobelia Cardinalis," *AJP* 11 (1840): 280–83.

"On the Power of Saccharine Substances in General, and Uncrystallizable Sugar in Particular, of Protecting the Solution of Protoiodide of Iron from Decomposition," *AJP* 12 (1840): 13–18.

"Description and Analysis of a Mineral[,] a Saline Incrustation, and a Bark, Brought from the Country Lying between Sante Fe, and the Headwaters of the Arkansas River," *AJP* 12 (1840): 108–14.

"On the Tartrate of Iron and Potassa," *AJP* 12 (1840): 188–92.

"On the Tartrate of Iron and Ammonia," *AJP* 12 (1840): 275–79.

"Solution of Iodide of Iron," *AJP* 12 (1840): 323.

"On Lobelina, the Active Principle of Lobelia Inflata, and on Some Other Proximate Principles of the Seed of that Plant," *AJP* 13 (1841): 1–10.

"Observations on Asarum Canadense, and Asarum Europoeum," *AJP* 13 (1841): 177–82.

"Observations on Blistering Plaster," *AJP* 13 (1842): 302–5.

"Remarks on Syrup of Wild Cherry Bark, and on Syrup of Valerian," *AJP* 14 (1842): 27–28 [with J. C. Turnpenny].

"Observations on Hydrated Peroxide of Iron, Demonstrative of its Decrease in Power, as an Antidote for Arsenious Acid, by Age, and Some Hints on the Methods of Preparing it," *AJP* 14 (1842): 29–37.

"Remarks on Some Pharmaceutical Preparations of Lobelia Inflata," *AJP* 14 (1842): 108–9.

"Observations on the Volatile Oil of Gaultheria Procumbens, Proving it to be a Hydracid Analogous to Saliculous Acid," *AJP* 14 (1842): 211–21.

"Observations on Extract of Rhatany," *AJP* 14 (1843): 267–72.

"Report on the Committee to whom was Referred the Paper of Thomas J. Husband, on Syrup of Sarsaparilla," *AJP* 15 (1843): 11–15.

"Elements of Chemistry, Including the Applications of the Science in the Arts by Thomas Graham [review]," *AJP* 15 (1843): 193–94.

"On the Naptha of Dr. Hastings," *AJP* 15 (1843): 194–95.

"Observations on the Volatile Oil of Betula Lenta, and on Gaultherin, A Substance which, by its Decomposition, Yields that Oil," *AJP* 15 (1844): 241–50.

"Report on a Root Found in Senega," *AJP* 15 (1844): 256–57 [with Augustine Duhamel].

"Notes on Valerianate of Zinc," *AJP* 17 (1845): 2–4.

"Observations on Monardin, a Peculiar Crystalline Substance, Derived from the Volatile Oil of Monarda Punctata," *AJP* 17 (1845): 86–89.

"Elementary Chemistry, Theoretical and Practical by George Fownes [review]," *AJP* 17 (1845): 166–69.

"Observations on Acetate of Zinc," *AJP* 17 (1846): 241–43.

"Remarks on the Preparations of Seneka," *AJP* 17 (1846): 247–52.

"Remarks on the Revision of the Pharmacopoeia," *AJP* 18 (1846): 1–9.

"Observations on a New Displacement Apparatus, Invented and Patented by Charles Augustus Smith, of Cincinnati," *AJP* 18 (1846): 98–103.

"On the Ethereal Extract of Cubebs," *AJP* 18 (1846): 167–69.

"On an Improved Drug Mill Invented by Beriah Swift," *AJP* 18 (1846): 255–56.

"On Sophistications of Sulphate of Quinia and Blue Mass," *AJP* 18 (1846): 267–69.

"Necrology [Augustine Duhamel]," *AJP* 18 (1846): 315–16.

"Remarks on the Reduction of Iron by Hydrogen," *AJP* 19 (1847): 11–16.

"Note on Cucumber Ointment," *AJP* 19 (1847): 16–17.

"On a Variety of Cinchona Bark," *AJP* 19 (1847): 178–82.

"On the Fluid Extracts of Rhubarb and Valerian," *AJP* 19 (1847): 182–87.

"Lecture, Introductory to the Course on Practical Pharmacy, Delivered at the Hall of the Philadelphia College of Pharmacy, October 25, 1847," *AJP* 19 (1847): 241–57.

"Elementary Chemistry, Theoretical and Practical by George Fownes [review]," *AJP* 19 (1847): 258.

"On the Anti-Septic and disinfecting Powers of Chloride of Zinc and Nitrate of Lead," *AJP* 19 (1847): 269–76.

"On the Decomposing Power of Water at High Temperatures, in a Scientific and Industrial Point of View, as Developed by R. A. Tighlman," *AJP* 20 (1848): 184–97.

"Concise Historical Sketch of the Progress of Pharmacy in Great Britain. Intended as an Introduction to the Pharmaceutical Journal by Jacob Bell, London, 1843, A," *AJP* 20 (1848): 265–79.

"Chemical Technology, or Chemistry Applied to the Arts and Manufactures by Dr. F. Knapp [review]," *AJP* 20 (1848): 280–87.

"Report to the College of Pharmacy on a Machine for Making Mercurial Pill Mass, Invented by J. W. W. Gordon of Baltimore," *AJP* 21 (1849): 6–12 [with T. P. James, William J. Jenks, Ambrose Smith, Edward Parrish, and John H. Ecky].

"Remarks on Oleo-Resinous Ethereal Extracts, Their Preparation, and the Advantages They Offer to the Medical Practitioner," *AJP* 21 (1849): 114–20.

"Observations on the Present State of Pharmacy in England," *AJP* 22 (1850): 1–18.

"On Hydrargyrum cum Creta Changed by Oxidation," *AJP* 22 (1850): 113–17.

"On American Narcotic and Other Extracts," *AJP* 22 (1850): 205–9.

"On the Importance of a More Conscientious Attention to Uniformity of Strength in the Preparation of Opium," *AJP* 22 (1850): 210–12.

"Observations on Copaiba," *AJP* 22 (1850): 289–96.

"On Some Pharmaceutical Preparations of Manganese," *AJP* 22 (1850): 297–303.

"Observations on an Oleo-Resin from Venezuela," *AJP* 23 (1851): 1–2.

"On the Means for Determining the Purity of Certain Chemicals and Drugs, and for Detecting Adulterations," *AJP* 23 (1851): 3–6.

"Remarks on Cod Liver Oil," *AJP* 23 (1851): 97–106.

"On a New Solvent for Extracting Cantharidin, and on the Existence of that Principle in Cantharis Vittata and Mylabris Cichorii," *AJP* 23 (1851): 124–26.

"On Solution of Citrate of Magnesia," *AJP* 23 (1851): 214–17.

"Remarks on Fluid Extracts of Cinchona," *AJP* 23 (1851): 218–19.

"Observations on the Sassy Bark of Western Africa and on the Tree Producing It," *AJP* 23 (1851): 301–12.

"On Liquor Ferri Nitratis U.S.P., 1850; and on a Formula for Syrup of Proto-Nitrate of Iron," *AJP* 23 (1851): 312–15.

"On Erythrophleum Juniciale (The Sassy Bark Tree of Cape Palmas)," *AJP* 24 (1852): 195–202.

"Extractum Lobeliae Fluidum," *AJP* 24 (1852): 207–8.

"On Hoffman's Anodyne Liquor," *AJP* 24 (1852): 213–18.

"Observations on the Volatility and Solubility of Cantharidin in View of the Most Eligible Pharmaceutical Treatment of Spanish Flies," *AJP* 24 (1852): 293–304.

"On Gelseminum Sempervirens or Yellow Jassamin," *AJP* 24 (1852): 307–10.

"Pharmaceutical Apparatus," *AJP* 24 (1852): 314–16.

"[Report of the Executive Committee] to the Convention of Pharmaceutists Assembled at Philadelphia, October 6th, 1852," *Proc APhA* 1 (1852): 7–12 [with Samuel M. Colcord and George D. Coggeshall].

Valedictory Address to the Graduates of the Philadelphia College of Pharmacy Delivered in Sansom Street Hall, March 18, 1852. Philadelphia: Merrihew and Thompson, 1852.

"On Fluid Extract of Rhubarb and Senna," *AJP* 25 (1853): 23–24.

"On Tannate of Alumina," *AJP* 25 (1853): 25–26.

"Description of Some Pharmaceutical Apparatus," *AJP* 25 (1853): 27–31.

"Note on Adulterated Powdered Socotrine Aloes," *AJP* 25 (1853): 99–101.

"Remarks on the Preparation of Hydrated Sequioxide of Iron as an Antidote, and on the Duty of the Pharmaceutist in Regard to It," *AJP* 25 (1853): 104–7.

"On Veratrum Viride," *AJP* 25 (1853): 109–13.

"On Syrup of Iodide of Iron and Manganese," *AJP* 25 (1853): 198–200.

"Emplastrum Extracti Aconiti Radicis," *AJP* 25 (1853): 202–3.

"Infusum Juniperi Compositum," *AJP* 25 (1853): 205–6.

"Linimentum Aconiti Radicis," *AJP* 25 (1853): 293–94.

"Remarks on Bogota and Other Cinchona Barks from New Grenada, in Their Relation to the Manufacture of Sulphate of Quinia," *AJP* 25 (1853): 306–12.

"On the Progress of Pharmacy in Great Britain," *AJP* 25 (1853): 385–98.

"On Fluid Extract of Gentian, and Some Remarks on Other Tonic Fluid Extracts," *AJP* 26 (1854): 28–30.

"On the Pharmacy of Cimicifuga," *AJP* 26 (1854): 106–7.

"Eclectic Pharmacy," *AJP* 26 (1854): 108–11.

"Additional Remarks on the Pharmacy of the Phosphates," *AJP* 26 (1854): 112–15.

"Pulvis Ferri: Iron by Hydrogen," *AJP* 26 (1854): 217–20.

"On the Incompatibility of Iodide of Potassium with Calomel and other Preparations of Mercury, and on a Simple Mode of Testing Pills, Powders, and other Medicines for the Presence of Mercury," *AJP* 26 (1854): 222–24.

"On the Preparation of the Syrups of Orange Peel, Tolu and Ginger, and on the Syrup and Fluid Extract of Vanilla," *AJP* 26 (1854): 298–301.

"Iodide of Sodium," *AJP* 26 (1854): 305.

"On the Production of Wine Brandy, and Tartar in the Valley of the Ohio," *AJP* 26 (1854): 411–14.

"On a New Variety of Flaxseed," *AJP* 26 (1854): 493–94.

"Remarks on Cupreous Soda Water, with Comments [by the Editor]," *AJP* 26 (1854): 496–99 [with Edwin R. Swann].

"[Report of the Committee on Education] to the Pharmaceutists of the United States," *Proc APhA* 3 (1854): 14–22 [with E. Parrish, D. Stewart, and J. Meakim].

"On the Volatile Oil of Erigeron Canadense," *AJP* 26 (1854): 502.

"Remarks on Gum Mezquite: A Variety of Gum Discovered by Dr. George G. Shumard, in Northern Texas," *AJP* 27 (1855): 14–17.

"On Benzoic Acid from Urine," *AJP* 27 (1855): 23–24.

"On the Culture and Manufacture of Castor Oil in Illinois and St. Louis," *AJP* 27 (1855): 99–103.

"On the Volatile Oil of Erigeron Philadelphicum," *AJP* 27 (1855): 105–6.

"On American Oil of Peppermint," *AJP* 27 (1855): 221–23.

"Remarks on Gum Mesquite," *AJP* 27 (1855): 223–25.

"Extractum Ignatiae Amarae Alcoholicum," *AJP* 27 (1855): 227–28.

"Extractum Sumbul Fluidum," *AJP* 27 (1855): 233–34.

"American Extract of Liquorice," *AJP* 27 (1855): 311–13.

"On the So-Called Cream Syrups for Mineral Water," *AJP* 27 (1855): 407–9.

"Aromatic Sysrup of Galls," *AJP* 27 (1855): 416.

"On Adulterated Oil of Aniseed," *AJP* 27 (1855): 513–14.

"Emplastrum Arnicae," *AJP* 27 (1855): 514.

"Report of the Committee on Standards for Drugs," *Pro APhA* 4 (1855): 15–16 [with George D. Coggeshall and Edward S. Wayne].

"Extractum Pruni Virginianae Fluidum," *AJP* 28 (1856): 21–23.

"On the Falsification of Strychnia with Oxalic Acid," *AJP* 28 (1856): 24–25.

"Remarks on a Carbonic Acid Water Apparatus," *AJP* 28 (1856): 103–6.

"Further Remarks on Fluid Extract of Wild Cherry Bark, and on the Alcoholic Extract and Wine of Wild Cherry Bark," *AJP* 28 (1856): 107–10.

"Test for the Presence of Alcohol in Chloroform," *AJP* 28 (1856): 213.

"Remarks on Flavoring Extracts," *AJP* 28 (1856): 215–17.

"A Few Hints Relative to the Collection of Some Indigenous Drugs," *AJP* 28 (1856): 411–13.

["Report of the Committee on Standards of Quality for Drug Examiners"], *Proc APhA* 5 (1856): 9 [with George D. Coggeshall and E. S. Wayne].

"[Report of the Committee on Text-books]," *Proc APhA* 5 (1856): 16–19 [with Charles Ellis and A. P. Sharp].

"Abstract of the Report of the Corresponding Secretary on the Statistics of Pharmacy and of the Drug Trade," *Proc APhA* 5 (1856): 77–79.

"On Valerianate of Ammonia," *AJP* 29 (1857): 27–28.

"Fluid Extract of Lupulin," *AJP* 29 (1857): 28–29.

"Report on the Progress of Pharmacy," *Proc APhA* 6 (1857): 50–79 [with Eugene Dupuy and James Cooke].

"Note on Collodion," *AJP* 29 (1857): 105–6.

"On a New Process for Making Liquor Ferri Nitratis," *AJP* 29 (1857): 306–7.

"On Syrup of Pyrophosphate of Iron," *AJP* 29 (1857): 404–5.

"Eau de Vichy: Artificial Vichy Water," *AJP* 29 (1857): 405–8.
"Remarks on Ergot," *Proc APhA* 6 (1857): 127–31.
"Remarks on the Medicinal Hypophosphites," *AJP* 30 (1858): 118–23.
"Valedictory Address to the Gradutates of the Philadelphia College of Pharmacy, Delivered March 11th, 1858, at the Musical Fund Hall," *AJP* 30 (1858): 193–205.
"Note on Compound Syrup of Hypophosphites," *AJP* 30 (1858): 226–27.
"On the Hypophosphites of Iron," *AJP* 30 (1858): 311–14.
"Does Nicotina Exist in Green Tobacco, or Is it a Result of Fermentation in the Curing Process," *AJP* 30 (1858): 502–8.
"Thoughts on 'Manufacturing Pharmacy' in Its Bearing on the Practice of Pharmacy, and the Character and Qualifications of Pharmaceutists," *AJP* 30 (1858): 514–17.
"Syllabus of a Course of Study, Intended as an Aid to Students of Pharmacy, Who Cannot Avail Themselves of Regular Instruction," *Proc APhA* 7 (1858): 103–68.
"Essay on Cerasus Serotina—The Wild Cherry Bark Tree of the United States, An," *Proc APhA* 7 (1858): 319–26.
"Citrate of Iron and Strychnia," *AJP* 31 (1859): 23–24.
"Remarks on Propylamin," *AJP* 31 (1859): 125–27.
"Note on Propylamin," *AJP* 31 (1859): 222–23.
"Note on the Kino of Eucalyptus Resinifera," *AJP* 31 (1859): 226–28.
"On Percolation or Displacement," *AJP* 31 (1859): 317–24.
"Remarks on Monsel's Persulphate of Iron," *AJP* 31 (1859): 403–7.
"Leaves of Wild Cherry, The" *AJP* 31 (1859): 423–24.
"Formulae for the Fluid Extracts in Reference to Their More General Adoption in the Next Pharmacopoeia," *AJP* 31 (1859): 530–49.
"Note on Commercial Chloric Ether," *AJP* 31 (1859): 553–54.
"Coumarin from Liatris Odoratissima," *AJP* 31 (1859): 556–57.
"Report . . . on the Progress of Pharmacy," *Proc APhA* 8 (1859): 49–114.
"Note on Kavaine and Methysticine," *AJP* 32 (1860): 133–35.
"Remarks on Polygalic Acid," *AJP* 32 (1860): 149–53.
"On Pills of Iodide of Iron," *AJP* 32 (1860): 201–3.
"On Syrup of Lactucarium and Some Other Syrups," *AJP* 32 (1860): 218–20.
"On Liquor Ferri Peracetatis," *AJP* 32 (1860): 309–10.
"Note on Japanese Wax," *AJP* 32 (1860): 311.
"On Species St. Germain," *AJP* 32 (1860): 312.
"Note on Fluid Extract of Wild Cherry Bark," *AJP* 32 (1860): 399–401.
"California Beer, or Yeast Plant," *AJP* 32 (1860): 409–10.
"On the Production of Atropia from American Grown Belladonna Root," *AJP* 32 (1860): 526–29.
"On the Preparation of Emplastrum Assafoetidae," *AJP* 32 (1860): 533–35.
"Prefatory Note," *Proc APhA* 9 (1860): xiii–xvi.
"What Are the Claims of Guaiac to Be Considered a Balsam," *Proc APhA* 9 (1860): 167–69.
"On Tinctura Arnicae," *AJP* 33 (1861): 11–12.
"Bitter Wine of Iron," *AJP* 33 (1861): 18–20.
"On Medicinal Aconitia and Its Substitutes," *AJP* 33 (1861): 100–104.

"Note on the Preparation of Chloroform, and of the Oil of Cloves," *AJP* 34 (1862): 25–28.

"On the Fluid Extract and the Syrup of Ipecacuanha," *AJP* 34 (1862): 28–29.

"On Sulphate of Anilin," *AJP* 34 (1862): 295–99.

"Remarks on Dialysis," *AJP* 34 (1862): 312–18.

"Horsely's Test for Strychnia," *AJP* 34 (1862): 400.

"On Liquid Rennet, or Rennet Wine," *AJP* 35 (1863): 36–38.

"American Pharmaceutical Association," *Drugg Circ* 7 (1863): 75.

"Notice: American Pharmaceutical Association," *AJP* 35 (1863): 279.

"American Pharmaceutical Association," *AJP* 35 (1863): 372.

"Note on Poisonings by Rhu Toxicodendron," *AJP* 35 (1863): 506.

"On the Relative Activity of American and European Aconite Root," *Proc APhA* 11 (1863): 196–200.

"On a Still for Apothecaries," *Proc APhA* 11 (1863): 207–8.

"On Fluid Extracts," *Proc APhA* 11 (1863): 222–48.

"On Economy in the Use of Alcohol in Pharmacy," *AJP* 36 (1864): 1–3.

"On the Relative Activity of American and European Aconite Root," *AJP* 36 (1864): 5–9.

"Note on Some Properties of Berberina," *AJP* 36 (1864): 10–12.

"On a Still for Apothecaries," *AJP* 36 (1864): 12–15.

"On the Sweet Gum Tree," *Proc APhA* 12 (1864): 222–23.

"Remarks on Dialysis," *Proc APhA* 12 (1864): 264–70.

"On a Test for the Resin of Cannabis Indica," *AJP* 37 (1865): 23–27.

"On the Odor of Commercial Tannic Acid," *AJP* 37 (1865): 53–54.

"Note on Caramania Gum," *AJP* 37 (1865): 105–7.

"Officinal Fluid Extracts, The," *AJP* 37 (1865): 181–83.

"Remarks on Some Preparations of the U.S. Pharmacopoeia," *AJP* 36 (1864): 209–12, 298–306, 393–406; *AJP* 37 (1865): 326–33.

"On Liquidambar Styraciflua and its Balsamic Resin," *AJP* 38 (1866): 33–38.

"Sulphocyanide of Mercury: Pharaoh's Serpents. Oleate of Soda and Soap Bubbles," *AJP* 38 (1866): 99–102.

"Note on Oleoresina Cubebae," *AJP* 38 (1866): 210–12.

"Note on Rectified Oil of Amber as a Remedy for Haemorrhoids," *AJP* 38 (1866): 217–18.

"On the Removal of Cincho-Tannic Acid from Liquid Preparations of Cinchona," *Proc APhA* 14 (1866): 223–24.

"Medical Purveying for the U.S. Army during the Late War," *AJP* 38 (1866): 271–74.

"Local Anaesthesia from the Cold of Rapid Evaporation and Spray-producing Instruments, for the 'Pulverization of Fluid,'" *AJP* 38 (1866): 289–93.

"Aubergier's Syrup of Lactucarium," *AJP* 38 (1866): 293–94.

"On Flavoring Extracts," *AJP* 38 (1866): 294–98.

"Note on Battley's Sedative Solution of Opium," *AJP* 38 (1866): 304–5.

"Essay on Sassafras Officinale, An," *AJP* 38 (1866): 481–92.

"Note on 'Alcoholized Iron,'" *AJP* 39 (1867): 11–12.

"On Extractum Colocynthidis Alcoholicum, U.S.P.," *AJP* 39 (1867): 15–20.

"Solution of Meconate of Morphia," *AJP* 39 (1867): 104–6.

"Note on Testing Glycerin for Sugar and Glucose," *AJP* 39 (1867): 109–11.

"Note on Narceia," *AJP* 39 (1867): 111–12.
"Note on 'Tinctura Lycoperdon,'" *AJP* 39 (1867): 113–14.
"Notes of Travel in Europe," *AJP* 40 (1868): 17–29.
"Gelseminum, *versus* Jasminum," *AJP* 40 (1868): 30.
"Visit to Vesuvius, A," *AJP* 40 (1868): 177–84.
"Botany in Its Bearings on Pharmacy—Botanical Gardens," *AJP* 40 (1868): 228–37.
"Obituary [of Stephen Procter]," *AJP* 40 (1868): 287.
"Note on a False Jalap," *AJP* 40 (1868): 389–92.
"On the New Act of Parliament Relative to Pharmacy and the Sale of Poisons," *AJP* 40 (1868): 397–400.
"Note on American Opium from Vermont," *AJP* 40 (1868): 513–16.
"Paris Exposition of 1867, The," *AJP* 40 (1868): 518–20; *AJP* 41 (1869): 8–13, 108–14.
"On Tinctura Physostigmae," *AJP* 40 (1868): 520–21.
"Vacuum Maceration: Duffield's Process for Fluid Extracts," *AJP* 41 (1869): 2–4.
"Additional Note on American Opium from Vermont," *AJP* 41 (1869): 23–26.
"Culture of Opium in the United States," *AJP* 41 (1869): 120–21.
"Labarraque's Wine of Quinium," *AJP* 41 (1869): 123–24.
"Note on Virginia Opium," *AJP* 41 (1869): 209–10.
"Vermont Opium Again," *AJP* 41 (1869): 217–18.
"On Titrated Opium Extract: Svapnia," *AJP* 41 (1869): 222–30.
"Pharmacopoeia of 1870: Shall Its Authority Be More Generally Respected?, The," *AJP* 41 (1869): 289–91.
"On Bicarbonate of Ammonia as a Pharmaceutical Preparation," *AJP* 41 (1869): 294–95.
"Proper Strength of Fluid Extracts, The," *AJP* 41 (1869): 295–96.
"'Sweet Quinine' What Is It?" *AJP* 41 (1869): 302–4.
"Carby's Prophylactic Fluid," *AJP* 41 (1869): 396–97.
"Castor Oil and Glycerin Pomatum," *AJP* 41 (1869): 398.
"Note on a Sample of So-Called Opium from Illinois," *AJP* 42 (1870): 68–69.
"Notice of M. Carre's Apparatus for Making Ice," *AJP* 42 (1870): 102–6.
"Assay of a Pure American Opium, from Poppies Grown at Hancock, Vermont, by Mr. C. M. Robbins," *AJP* 42 (1870): 124–27.
"Note on Sulpho-Carbolic Acid and the Sulpho-Carbolates," *AJP* 42 (1870): 131–36.
"Note on Cod-Liver Oil and Other Products from Portsmouth, N.H.," *AJP* 42 (1870): 214–15.
"Abies Canadensis," *Proc APhA* 18 (1870): 134–37.
"Carre's Apparatus for Making Ice," *AJP* 43 (1871): 1–3.
"Little Side-Talk with the 'Freshmen' in Pharmacy, A," *AJP* 43 (1871): 7–10.
"On a Morphiometric Process for the Pharmacopoeia," *AJP* 43 (1871): 65–67.
"Can Practical Pharmacy be Taught Effectively by Lectures?" *AJP* 43 (1871): 97–100.
"Pharmaceutical Titles," *AJP* 43 (1871): 157–58.
"Elias Durand [obit.]," *AJP* 45 (1873): 513–14.
"Suggestions to Beginners in Pharmacy," *Proc APhA* 21 (1873): 523–32.
"On Cucumber Ointment," *Proc APhA* 21 (1873): 601–3.

"On Orange-Colored Glass as a Means of Protecting Volatile Oils," *Proc APhA* 21 (1873): 629–30.

"Editorial Department" with Joseph Carson—*AJP* 20 (1848): 349–52; *AJP* 21 (1849): 93–96, 191–92, 287–88, 382–84; *AJP* 22 (1850): 189–92, 279–88, 377–84.

"Editorial Department"—*AJP* 23 (1851): 91–96, 186–95, 286–301, 385–92; *AJP* 24 (1852): 85–98, 183–91, 281–92, 381–90; *AJP* 25 (1853): 86–96, 187–92, 282–88, 369–84, 473–80, 572–78; *AJP* 26 (1854): 89–96, 169–92, 286–88, 378–84, 473–80, 567–76; *AJP* 27 (1855): 85–94, 185–92, 284–88, 379–84, 477–80, 568–76; *AJP* 28 (1856): 88–94, 187–92, 285–88, 377–84, 478–80, 573–76; *AJP* 29 (1857): 86–94, 186–92, 283–88, 379–84, 477–80, 572–76; *AJP* 30 (1858): 88–94, 185–92, 280–88, 381–84, 464–76, 574–82; *AJP* 31 (1859): 86–94, 186–92, 284–88, 383–94, 484–90, 583–86; *AJP* 32 (1860): 89–94, 187–92, 283–88, 376–84, 468–80, 571–76; *AJP* 33 (1861): 89–94, 185–92, 287–88, 382–84, 476–80, 575–76; *AJP* 34 (1862): 93–94, 190–92, 286–88, 378–84, 491–94, 565–75; *AJP* 35 (1863): 90–94, 187–92, 282–88, 373–84, 476–80, 569–76; *AJP* 36 (1864): 91–94, 181–92, 284–88, 380–84, 461–64, 538–44; *AJP* 37 (1865): 74–78, 156–60, 234–40, 314–20, 396–400, 492–96; *AJP* 38 (1866): 88–94, 186–92, 278–88, 379–84, 475–80, 569–76; *AJP* 39 (1867): 87–94, 187–92, 283–88, 565–78; *AJP* 40 (1868): 87–94, 186–92, 281–88, 378–84, 467–80, 567–76; *AJP* 41 (1869): 81–94, 182–92, 280–88, 371–84, 469–80, 569–76; *AJP* 42 (1870): 81–93, 179–92, 280–88, 376–84, 471–80, 568–76; *AJP* 43 (1871): 42–46, 92–96, 139–44, 185–92.

"Varieties"—*AJP* 24 (1852): 169–80; *AJP* 26 (1854): 272–82, 365–77, 460–72, 554–64; *AJP* 27 (1855): 74–84, 176–85, 273–78, 371–78, 467–76, 558–66; *AJP* 28 (1856): 78–87, 175–87, 267–73, 365–74, 466–76, 561–69; *AJP* 29 (1857): 82–85, 177–85, 371–79, 465–71, 561–68; *AJP* 30 (1858): 81–88, 175–84, 274–77, 374–79, 461–63, 569–70; *AJP* 31 (1859): 279–80, 478–83 [with John Maisch], 579–81; *AJP* 32 (1860): 180–86, 275–76, 466–67, 562–64; *AJP* 33 (1861): 377–81, 473–75, 569–72; *AJP* 34 (1862): 89–93, 185–89, 377; *AJP* 35 (1863): 182–87, 273–78, 466–76; *AJP* 36 (1864): 378–80; *AJP* 37 (1865): 390–95, 491–92; *AJP* 38 (1866): 86–88, 180–85, 375–78, 472–75, 561–68; *AJP* 39 (1867): 84–87, 181–87; *AJP* 40 (1868): 81–84, 184–85, 273–77; *AJP* 42 (1870): 467–71.

"Gleanings"—*AJP* 25 (1853): 211–14; *AJP* 26 (1854): 9–17, 220–22, 301–4, 504–5; *AJP* 27 (1855): 17–22, 106–14, 229–33, 313–19; *AJP* 28 (1856): 114–16, 218–22, 524–25; *AJP* 29 (1857): 310–15, 316–20, 398–404; *AJP* 30 (1858): 33–34, 124–28, 397–405; *AJP* 31 (1859): 229–32, 414–17; *AJP* 32 (1860): 220–25, 316–19; *AJP* 33 (1861): 212–15, 221–24, 319–21, 411–14, 415–16, 500–505; *AJP* 34 (1862): 139–41, 318–26, 395–97, 506–12; *AJP* 35 (1863): 223–27, 312–16, 507–12; *AJP* 36 (1864): 16–21, 113–14, 213–14, 312–14, 417–20; *AJP* 37 (1865): 253–55, 333–37; *AJP* 38 (1866): 298–304, 504–7; *AJP* 39 (1867): 219–20, 330–34, 394–96, 521–24; *AJP* 41 (1869): 14–19, 27–32, 124–28, 204–8; *AJP* 42 (1870): 509–13.

"Pharmaceutical Notices"—*AJP* 13 (1841): 183–90; *AJP* 19 (1847): 93–97, 259–65; *AJP* 20 (1848): 85–92; *AJP* 22 (1850): 108–13; *AJP* 24 (1852): 219–25; *AJP* 25 (1853): 408–11; *AJP* 29 (1857): 108–11; *AJP* 34 (1862): 136–39, 208–11; *AJP* 41 (1869): 388–92.

Manuscript Materials

American Pharmaceutical Association, Archives, Washington, D.C. Student notebook of C. Lewis Diehl of classes at Philadelphia College of Pharmacy, 1859.

———. Prescription book of William Procter, Jr., January–March 1847.

College of Physicians of Philadelphia, Philadelphia, Penn. George Bacon Wood Papers.

Historical Society of Pennsylvania, Manuscripts Department, Philadelphia, Penn. Lea and Febiger Collection.

———. Minutes of the Philadelphia Drug Exchange, 1861–1921.

Kremers Reference Files, University of Wisconsin-Madison School of Pharmacy, Madison, Wisc. File A2: Procter, William. Correspondence of William Procter, Jr., to Albert Ebert.

———. Student notebook of Frederick B. Power of classes at the Philadelphia College of Pharmacy, 1874. (Photocopy of original in holdings of the Wellcome Institute, London, England.)

Philadelphia City Hall, Philadelphia, Penn. Registry of Wills. Wills of William Procter, Jr., Margaretta Procter, Catherine Procter, and Charles Ellis.

Philadelphia City Archive, City Hall Annex, Philadelphia, Penn. Registration of Deaths in the City of Philadelphia, 1874.

———. Purchase Journal, Drugs and Laboratory Supplies, of the Guardians of the Poor [Almshouse], hospital, 1860–77.

Philadelphia College of Pharmacy and Science, Philadelphia, Penn. "Certificates presented to Professor Wm. Procter, Jr., 1839 to 1873." Scrapbook in rare book collection, Joseph England Library.

———. Journal of the Committee, on the Revision of the U.S. Pharmacopoeia, of the Philadelphia College of Pharmacy [c. 1840]. Vault in dean's office.

———. [Minutes of the Committee on the Revision of the United States Pharmacopoeia of the Philadelphia College of Pharmacy, 1847], mss. bound with Journal of the Committee . . . [c. 1840].

———. Minute Book of the Trustees of the Philada. College of Apothecaries [Pharmacy]. [Vol. 1, 1821–40.] Rare book collection.

———. Minute Book of the Philadelphia College of Pharmacy, 1842–69. Rare book collection.

———. Minute Book of the Philadelphia College of Pharmacy, 1870–87. Rare book collection.

———. "Report of the Committee of Revision of the U.S. Pharmacopoeia of the Philadelphia College of Pharmacy, 1840." 2 vols. Vault in dean's office.

———. "Report of the Committee Appointed by the Philadelphia College of Pharmacy to Revise the Pharmacopoeia of U.S., 1850." Vault in dean's office.

———. "Report of the Committee Appointed by the Philadelphia College of Pharmacy to Revise the Pharmacopoeia of U.S., 1860." Vault in dean's office.

———. "Minutes of the Committee of Revision of the United States Pharmacopoeia, 1863." Vault in dean's office. [Note: these are the minutes of the revision committee empowered by the National Medical Convention, not by the College of Pharmacy.]

―――. "Minutes of the Committees of the Philadelphia College of Pharmacy Appointed in Relation to Erection of a New Building [1867]." Rare book collection.

―――. "Minutes of the Committee on the World's Fair Deposite, 1861–2." Rare book collection.

Smithsonian Institution Archives, Washington, D.C. Ingoing and outgoing correspondence of Spencer Baird.

Published Materials and Dissertations

A. M'Elroy's Philadelphia Directory for 1837. Philadelphia: M'Elroy, 1837.

"Abstract from the Minutes of the Philadelphia College of Pharmacy," *AJP* 22 (1850): 184–88; *AJP* 25 (1853): 561–64; *AJP* 34 (1862): 274–85, 560–64; *AJP* 35 (1863): 280–81, 566–69; *AJP* 36 (1864): 537–38; *AJP* 38 (1866): 568–69.

Account of the New-York Hospital, An. New York: Collins, 1811.

"Annual Commencement of the Philadelphia College of Pharmacy," *AJP* 38 (1866): 274–76.

Anonymous. "Apothecaries," *Boston Medical and Surgical Journal* 11 (1834–35): 261.

Archibald, Henry C. "On Campbell's Process for the Manufacture of Fluid Extracts," *AJP* 42 (1870): 117–20.

Attfield, John. *Chemistry: General, Medical, and Pharmaceutical.* Philadelphia: Henry Lea, 1879.

―――. "Correspondence," *AJP* 78 (1906): 448.

Attfield, Professor [John]. "Observations on Ferric Hydrate," *AJP* 40 (1868): 454–56.

Ayer, James C. "The Alkaloids and Proximate Principles," *AJP* 25 (1853): 407–8.

Bache, Franklin. "Address Delivered before the Graduates of the Philadelphia College of Pharmacy, April 27, 1835," *AJP* 7 (1835): 89–104.

―――. "On the Advantages That Would Accrue to English and American Pharmacy by the Adoption of a Single Uniform Pharmacopoeia for the British Empire," *AJP* 27 (1855): 11–13.

―――. "Note on the Nomenclature of Salts," *AJP* 27 (1855): 213–16.

Bailey, M. J. "Communication [of the New York Examiner of Drugs]," *Proc APhA* 6 (1857): 100–104.

Bakes, William C. "On Baume Tranquille," *AJP* 34 (1862): 22–23.

Barr, John. "Extract of Cod Liver," *AJP* 38 (1866): 139–43.

Beatson, James. "On the Spontaneous Decomposition of Pyroxylin (Gun Cotton)," *AJP* 25 (1853): 19–22.

B[edford], P. W. "Obituary Record of Professor William Procter, Jr.," *Drugg Circ* 18 (1874): 68.

Bell, Jacob. "The Relative Positions of Medical and Pharmaceutical Reform at the Present Time," *Pharmaceutical Journal* 10 (1850): 55.

Bell, Whitfield J. *John Morgan: Continental Doctor.* Philadelphia: Univ. of Pennsylvania Press, 1965.

Benger, F. Baden. "The Apprenticeship and Early Training of Pharmacists [in England]," *AJP* 42 (1870): 543–46.

Benjamin, Philip S. *The Philadelphia Quakers in the Industrial Age*. Philadelphia: Temple Univ. Press, 1976.

Beringer, George M. "Tribute to the Memory of Professor Procter," *AJP* 76 (1905): 38–39.

Berman, Alex. "The Impact of the Nineteenth Century Botanico-Medical Movement on American Pharmacy and Medicine." Ph.D. diss., University of Wisconsin, 1954.

———. "A Striving for Scientific Respectability: Some American Botanics and the Nineteenth-Century Plant Materia Medica," *Bulletin of the History of Medicine* 30 (1956): 7–31.

B[igelow], J[acob]. "[Review of] The Pharmacopoeia of the United States of America . . . 1830," *American Journal of Medical Sciences* 8 (1831): 149.

Blochman, Lawrence G. *Doctor Squibb*. New York: Simon and Schuster, 1958.

Boettger, Rudolph. "On the Application of Starch Sugar as a Reducing Agent for Chloride of Silver," *AJP* 30 (1858): 537–38.

Boorstin, Daniel. *The Americans: The National Experience*. New York: Vintage Books, 1965.

Boring, Edwin M. "Tribute to Memory of Professor Procter," *AJP* 76 (1905): 41.

Brewer, W. A. "Reminiscences of an old pharmacist #2," *Pharmaceutical Record* 4 (1884): 232–33.

———. "Reminiscences of an old pharmacist #7," *Pharmaceutical Record* 4 (1884): 348–49.

———. "Reminiscences of an old pharmacist #10," *Pharmaceutical Record* 4 (1884): 442–43.

[Bridgman, Henry]. "Secret Remedies," *Drugg Circ* 1 (1857): 75.

———. "Wine and its Substitutes," *Drugg Circ* 1 (1857): 89.

———. "The Close of Our First Volume," *Drugg Circ* 1 (1857): 156.

Brief Account of the New-York Hospital, A. New York: Isaac Collins and Son, 1804.

Brock, W. R. *The United States, 1789–1890*. Ithaca, N.Y.: Cornell Univ. Press, 1975.

Bruchey, Stuart W., ed. *Small Business in American Life*. New York, Columbia Univ. Press, 1980.

Bullock, Charles. "Note on Chlorodyne," *AJP* 37 (1865): 17–19.

———. "Memoir of Prof. William Procter, Jr.," *AJP* 46 (1874): 512–33.

Bureau of Labor Statistics. *Historical Statistics of the United States, 1789–1945*. Washington, D.C.: GPO, 1949.

Burrow, James G. *AMA: Voice of American Medicine*. Baltimore: Johns Hopkins Univ. Press, 1963.

Bylaws and Regulations . . . of the New-York Hospital. New York: Mahlon Day, 1820.

C., J. "The Pharmacopoeia of the United States of America. Fourth Decennial Revision . . . [review]," *American Journal of the Medical Sciences* 46 (1863): 433.

C., W. P. "The Physician's Prescription. To whom Does It Belong?" *AJP* 38 (1866): 204–6.

Caldwell, A. "The Percentage System," *AJP* 42 (1870): 312–13.

Campbell, Samuel. "On a New and Simple Process for Fluid Extracts, by Which any Drug May Be Exhausted by Percolation and without Heat," *AJP* 41 (1869): 385–88.

————. "A Supplement to Campbell's Method of Percolation for Fluid Extracts," *AJP* 42 (1870): 17–22.

Carney, Charles T. "Report on the Revision of the Pharmacopoeia," *Proc APhA* 7 (1858): 177–83.

————. "Paraffin—Its Substitution for Wax in Cerates," *Proc APhA* 9 (1860): 163–65.

Carr-Saunders, A. M., and P. A. Wilson. *The Professions*. Oxford: Clarendon Press, 1933.

Carson, Gerald. *The Old Country Store*. New York: Oxford Univ. Press, 1954.

Carson, Joseph. "Address Delivered to the Graduates of the Philadelphia College of Pharmacy, April 23d, 1839," *AJP* 11 (1839): 89–103.

C[arson], J[oseph]. "Editorial Department," *AJP* 20 (1848): 253–56.

Cassedy, James. "The Flourishing and Character of Early American Medical Journalism, 1797–1860," *Journal of the History of Medicine and Allied Health Sciences* 38 (1983): 135–50.

Catalogue of Blanchard and Lea's Medical and Surgical Publications. Philadelphia: Blanchard and Lea, 1853.

"Catalogue of the Class of the Philadelphia College of Pharmacy," *AJP* 27 (1855): 95–96; *AJP* 28 (1856): 95–96; *AJP* 29 (1857): 95–96; *AJP* 30 (1858): 95–96; *AJP* 31 (1859): 95–96; *AJP* 32 (1860): 95–96; *AJP* 33 (1861): 95–96; *AJP* 34 (1862): 95–96; *AJP* 35 (1863): 95–96; *AJP* 36 (1864): 95–96; *AJP* 37 (1865): 79–80; *AJP* 39 (1867): 94–96; *AJP* 40 (1868): 94–96; *AJP* 41 (1869): 94–96; *AJP* 42 (1870): 94–96; *AJP* 43 (1871): 46–48; *AJP* 44 (1872): 46–48; *AJP* 45 (1873): 45–48.

Chapman, W. B. "On the Uncertainty of the Composition of Pharmaceutical Preparations, and the Most Eligible Form of Medicines for Administration," *AJP* 26 (1854): 154–59.

"Charter and Laws of the Philadelphia College of Pharmacy," *AJP* 28 (1856): 278–84.

"Chemistry of Aniline Colors, The," *AJP* 35 (1863): 347–48.

"Circular of the New York College of Pharmacy Relative to the Adulteration of Drugs," *AJP* 19 (1847): 305–9.

Clacius, C. E. "Sophistications," *AJP* 43 (1871): 317–19.

"Code of Ethics Adopted by the Philadelphia College of Pharmacy, A," *AJP* 20 (1848): 148–51.

"Code of Ethics of the Maryland College of Pharmacy," *AJP* 28 (1856): 374–76.

Coggeshall, George D. "An Address Delivered to the Graduates of the New York College of Pharmacy, Nov. 20th, 1845," *AJP* 18 (1846): 244–47.

————. "Address Delivered to the Graduates of the College of Pharmacy of the City of New York, March 16, 1854," *AJP* 26 (1854): 201–7.

Coggeshall, George D., James Aspinwall, and John Meakim. "Report of the New York College of Pharmacy on the Statistics of that City," *Proc APhA* 2 (1853): 35–36.

Colcord, Samuel M. "Report of the Massachusetts College of Pharmacy," *Proc APhA* 2 (1853): 24–26.

————. "Pharmaceutical Colleges and Associations: Massachusetts College of Pharmacy," *AJP* 46 (1874): 135–37.

Cole, Arthur Charles. *The Irrepressible Conflict, 1850–1865.* New York: Mac-millan, 1934.

"Commencement of the Philadelphia College of Pharmacy," *Drugg Circ* 5 (1861): 85.

"Convention for the Fifth Decennial Revision of the Pharmacopoeia of the United States, The," *AJP* 42 (1870): 289–96.

Cowen, David L. "Pharmacy in the Curriculum of American Medical Schools," *Pharmacy in History* 20 (1978): 17–21.

———. "Materia Medica and Pharmacology." In *The Education of American Physicians: Historical Essays,* edited by Ronald L. Numbers. Berkeley: Univ. of California Press, 1980.

Cowen, David L., Louis King, and Nicholas G. Lordi. "Nineteenth Century Drug Therapy: Computer Analysis of the 1854 Prescription File of a Burlington [New Jersey] Pharmacy," *Journal of the Medical Society of New Jersey* 78 (1981): 758–61.

Coxe, Edward Jenner. "Remarks on the Preparation of Hydrargyrum cum Creta," *AJP* 22 (1850): 316–18.

Crew, B. J. "On a Substitute for Tar Beer," *AJP* 27 (1855): 13–14.

Cummings, H. T. "Ice: Its Collection, Storage and Distribution," *AJP* 40 (1868): 211–23.

Curti, Merle E. *The Growth of American Thought.* 3d ed. New York: Harper and Row, 1964.

Curtman, Charles O. "Pharmaceutical Still," *AJP* 41 (1869): 197–200.

Dana, J. D. "On the Relations of Death to Life in Nature," *AJP* 35 (1863): 129–33.

Daniels, George H. *American Science in the Age of Jackson.* New York: Columbia Univ. Press, 1968.

Daniels, George H., et al. *Science in American Society.* New York: Knopf, 1971.

"Death of Professor William Proctor [*sic*], Jun.," *The Chemist and Druggist,* 15 (1874): 77.

Desilver's Philadelphia Directory and Stranger's Guide for 1837. Philadelphia: Robert Desilver, 1837.

"Died: Procter," (Philadelphia) *Public Ledger,* vol. 76, no. 120 (February 11, 1874), p. 2.

"Died . . . Procter," *Friends' Intelligencer* 30 (1874): 825.

Diehl, C. Lewis. "On Fractional Percolation," *AJP* 41 (1869): 337–45.

———. "Report on the Progress of Pharmacy," *Proc APhA* 22 (1874): 25–303.

Dingwall, Robert. "Accomplishing Profession," *Sociological Review* 24 (1976): 331–35.

Dollfus, Armand. "On the Sensibility of the Reaction of Saliculous and Salicylic Acids on the Sesquioxide of Iron," *AJP* 26 (1854): 65–66.

Donald, David Herbert. *Liberty and Union.* Boston: Little, Brown, 1978.

"Draft of a Proposed Law to Regulate the Practice of Pharmacy and the Sale of Poisons," *AJP* 41 (1869): 562–69 [written by John M. Maisch].

Duhamel, Augustine. "Pharmaceutical Notices," *AJP* 9 (1837): 291–96.

———. "Boullay's Filter and System of Displacement, with Observations Drawn from Experience," *AJP* 10 (1838): 1–17.

————. "Memoir upon the Detection of Arsenic Forming a Toxicological Compendium as far as regards this Substance," *AJP* 12 (1840): 279–97.

Duhamel, Augustine, and William Procter, Jr. "Observations on the Method of Displacement," *AJP* 11 (1839): 189–201.

————. "Report on a Root Found in Senega," *AJP* 15 (1843): 256–57.

Duhamel, Augustine, John H. Ecky, and Wm. Procter, Jr. "Report on Catechu," *AJP* 16 (1844): 164–66.

DuMez, A. G. "Procter, William." In *Dictionary of American Biography,* edited by Dumas Malone. New York: Scribner's Sons, 1935.

————. "Extraction." In *American Pharmacy,* edited by Rufus Lyman. Philadelphia: J. B. Lippincott, 1945.

Eatwell, W. C. B. "Observations on the Cultivation of the Poppy and the Manufacture of Opium in British India," *AJP* 24 (1852): 118–33.

Ebert, Albert E. "Pharmaceutical Notes," *AJP* 39 (1867): 107–9; *AJP* 42 (1870): 540–42.

————. "The Father of American Pharmacy: Professor William Procter, Jr.," *Proc APhA* 50 (1902): 153–56.

Ebert, Myrl. "The Rise and Development of the American Medical Periodical, 1797–1850," *Bulletin of the Medical Librarians Association* 40 (1952): 243–76.

Ehrman, John W. "Notes on Aromatic Sulphuric Acid and Confection of Senna," *AJP* 43 (1871): 122–25.

"Eighth Annual Meeting of the American Pharmaceutical Association, The," *Drugg Circ* 3 (1859): 222–24.

Ekirch, Arthur Alphonse. *The Idea of Progress in America, 1815–1860.* New York: Peter Smith, 1951.

Ellis, Evan T. "Tribute to the Memory of Professor Procter," *AJP* 76 (1905): 41.

Engelmann, Geo. J. "Chloral—A New Anaesthetic," *AJP* 41 (1869): 447–48.

England, Joseph W., ed. *First Century of the Philadelphia College of Pharmacy, 1821–1921.* Philadelphia: Philadelphia College of Pharmacy, 1922.

Estes, J. Worth. "John Jones's *Mysteries of Opium Reveale'd* (1701): Key to Historical Opiates," *Journal of the History of Medicine and Allied Sciences* 34 (1979): 200–209.

"Exhibition Connected with the 16th Annual Meeting of the American Pharmaceutical Association," *AJP* 40 (1868): 376.

"Extract from the Minutes of the Board of Trustees," *AJP* 18 (1846): 314.

"Extracts from the Minutes of the Pharmaceutical Meetings," *AJP* 18 (1846): 306–12.

Faber, John. "On Pharmacy in the United States," *AJP* 41 (1869): 398–404.

"First Session—Monday Afternoon, Sept. 16, 1901," *Proc APhA* 49 (1901): 1–26.

Fish, Carl Russell. *The Rise of the Common Man: 1830–1850.* New York: Macmillan, 1927.

Fisher, William R. "A Brief Sketch of the Progress and Present State of Pharmacy in the United States of America," *AJP* 9 (1837): 271–79.

Fishwick, Marshall. *The Hero, American Style.* New York: David McKay, 1969.

Freedley, Edwin T. *Leading Pursuits and Leading Men.* Philadelphia: Edward Young, 1856.

Freidson, Eliot. "The Theory of Professions: State of the Art." In *The Sociology of the Professions,* edited by R. Dingwall and P. Lewis. New York: St. Martin's Press, 1983.

French, Howard B. "Tribute to the Memory of Professor Procter," *AJP* 77 (1905): 35–36.

Frezenius, C. R. "Deportment of the Most Important Medicinal Alkaloids with Reagents, and a Systematic Method of Effecting the Detecting of these Substances," *AJP* 38 (1866): 546–54; *AJP* 39 (1867): 27–31.

Galen. "Philadelphia Correspondence of the Druggists Circular," *Drugg Circ* 1 (1857): 137.

Garrigues, S. S. "Minutes of the Pharmaceutical Meetings [of the Philadelphia College of Pharmacy]," *AJP* 31 (1859): 84–86.

Gibson, Melvin R. "Scope [of Pharmacy]." In *Remington's Pharmaceutical Sciences.* Easton, Penn.: Mack Publishing, 1970.

Gildemeister, E. *The Volatile Oils.* 2d ed. Translated by Edward Kremers. New York: John Wiley, 1922.

Gildemeister, E., and F. Hoffman. *The Volatile Oils.* Translated by Edward Kremers. Milwaukee: Pharmaceutical Review Publishing, 1900.

Goodrich, Chauncey A., revisor. *An American Dictionary of the English Language,* by Noah Webster. Springfield, Mass.: Merriam, 1854.

Grahame, Israel J. "Acetum Opii or Black Drop," *AJP* 30 (1858): 527–30.

———. "The Process of Percolation or Displacement: Its History and Application in Pharmacy," *Proc APhA* 7 (1858): 285–94.

Griffith, Ivor. "William Procter, Jr.," *AJP* 111 (1939): 44–47.

———. "William Procter, Jr.," *AJP* 113 (1941): 329–36.

———. "Wm. Procter, Jr.—Father of American Pharmacy," *AJP* 115 (1943): 406–16.

Gulick, Luther H. "On the Materia Medica of the Sandwich Islands," *AJP* 27 (1855): 234–40.

Guralnick, Stanley M. "The American Scientist in Higher Education, 1820–1910." In *The Sciences in the American Context: New Perspectives,* edited by Nathan Reingold. Washington, D.C.: Smithsonian Institution Press, 1979.

Guthrie, C. B., et al. "Report of a Committee to Consider and Report on the Subject of Home Adulterations," *AJP* 28 (1856): 126–28.

———. "Report of the Committee on Weights and Measures," *Proc APhA* 7 (1858): 169–71.

Guthrie, James. "Circular of Instructions to the Special Examiners of Drugs," *AJP* 25 (1853): 301–4.

Hagner, Charles V. "On Drug Grinding," *AJP* 23 (1851): 197–206.

Hammond, W. A. "Prohibition of Calomel and Tartar Emetic in the Army," *AJP* 35 (1863): 329–30.

Hanbury, Daniel. "Letter to the Editor on the Botanical Source of Balsam of Peru," *AJP* 32 (1860): 411–12.

———. "Note on the Use of Balsam of Peru in the Roman Catholic Church," *AJP* 33 (1861): 262–64.

———. "Note on Cassia Moschata," *AJP* 36 (1864): 80–83.

———. "On the Manufacture of Balsam of Peru," *AJP* 36 (1864): 145–51.

———. "On Pharmaceutical Herbaria," *AJP* 38 (1866): 334–37.

————. "On a Species of Ipomoea, Affording Tampico Jalap," *AJP* 42 (1870): 330–33.

Hancock, John F. "Pharmaceutical Colleges and Associations: Maryland College of Pharmacy," *AJP* 46 (1874): 138–39.

————. "William Procter, Jr., the Father of American Pharmacy," *AJP* 77 (1905): 13–20.

Hare, Robert. "Criticisms and Suggestions Respecting Nomenclature," *AJP* 9 (1837): 1–16.

Harrop, Joseph. "Improper Use of Titles," *AJP* 41 (1869): 221–22.

————. "On Syrup of Citric Acid of the Pharmacopoeia," *AJP* 41 (1869): 393.

Heintz, P. "On the Composition of Butter," *AJP* 26 (1854): 150–52.

Heintzelman, Joseph A. "Syrupus Stillingiae Compositus," *AJP* 34 (1862): 206–8.

Heydenreich, F. Victor. "On Capsicum Annuum," *AJP* 30 (1858): 296–301.

Higby, Gregory J. "*Practical Pharmacy* by Mohr, Redwood, and Procter: An International Pharmacy Textbook," *Pharmacy in History* 26 (1984): 97–102.

————. "Professionalism and the Nineteenth-Century American Pharmacist," *Pharmacy in History* 28 (1986): 115–24.

————, ed. *One Hundred Years of the National Formulary.* Madison, Wisc.: American Institute of the History of Pharmacy, 1989.

Hodgson, William, Jr. "Notes on Falsifications and Adulterations," *AJP* 9 (1837): 17–20.

Hoffman, Frederick. "Gleanings from German Journals," *AJP* 42 (1870): 394–401.

————. "The Application of the Microscope in Pharmacy and the Drug Trade," *Drugg Circ* 18 (1874): 57–58.

————. "A Century of American Pharmaceutical Literature and Journalism," *American Druggist and Pharmaceutical Record* 36 (1900): 159–65.

————. "Memorials to American Pharmacists," *AJP* 73 (1901): 83–86.

Hoffman, F., James Shinn, and George W. Sloan. "Recollections and Reminiscences of Prof. Wm. Procter, Jr.," *AJP* 72 (1900): 485–89.

Hooker, William, and Daniel Hanbury. "Botanical and Pharmacological Inquiries and Desiderata," *AJP* 32 (1860): 49–58.

Howard, J. E. "Microscopical Researches on the Alkaloids, as Existing in Cinchona Bark," *AJP* 37 (1865): 346–50.

Hudson, Robert P. "The Biography of Disease: Lessons from Chlorosis," *Bulletin of the History of Medicine* 51 (1977): 449–65.

Hughes, Everett C. *The Sociological Eye.* Chicago: Aldine-Atherton, 1971.

Hunt, T. Sterry. "On the History of Petroleum or Rock Oil," *AJP* 34 (1862): 527–41.

"Hydrargyri Iodidum Rubrum," *AJP* 24 (1852): 113.

Ince, Joseph. "Scheele and His Discoveries," *AJP* 36 (1864): 49–59.

————. "Obituary. Emeritus Professor Redwood," *Pharmaceutical Journal* 51 (1891–92): 763–66.

"Incompatibility of Oxide of Silver with Grape Sugar," *AJP* 30 (1858): 405–6.

"Introduction," *Journal of the Philadelphia College of Pharmacy* 1 (1825): 1–3.

Jackson, Samuel. "Case of Supposed Poisoning with Arsenic," *American Journal of Medical Sciences* 5 (1829): 237–43.

————. "On Belladonna in Pertussis," *American Journal of Medical Sciences* 14 (1834–35): 368.

"James W. Simes, Druggist and Chemist," (Philadelphia) *Public Ledger,* April 6, 1836, p. 1.

Jones, W. Ranstead. "Chalk Mixture," *AJP* 42 (1870): 393.

Kaufman, Martin. "American Medical Education." In *The Education of American Physicians: Historical Essays,* edited by Ronald L. Numbers. Berkeley: Univ. of California Press, 1980.

Kemp, George. "On a Direct Method of Obtaining Iodine from Certain Species of Sea-weed," *AJP* 22 (1850): 338–48.

Kennedy, George W. "Note on Campbell's Process for Fluid Extract," *AJP* 42 (1870): 62–63.

————. "Tincture of Nux Vomica," *AJP* 42 (1870): 392–93.

Kennedy, George W., and John M. Maisch. "Prefatory Notice," *Proc APhA* 22 (1874): 21–22.

Kiersted, H. T. "American Pharmaceutical Association: Notice of Postponement," *AJP* 33 (1861): 377.

————. "American Pharmaceutical Association—Notice," *AJP* 34 (1862): 378.

Kilmer, F. B., and Charles D. Deshler. "Drug Clerks One Hundred Years Ago," *Journal of the American Pharmaceutical Association* 18 (1929): 711–22.

King, James T. "On Mr. Campbell's Process for Preparing Fluid Extracts," *AJP* 42 (1870): 29–30.

[Kraemer, Henry]. "Editorial, Charles Rice," *AJP* 73 (1901): 303–11.

————. "Editorial, Memorials," *AJP* 74 (1902): 553–55.

————. "Editorial: The Procter Monument," *AJP* 77 (1905): 32–35.

————. "The Procter Memorial," *AJP* 78 (1906): 420–22.

Krummeck, Jacob. "Remarks on 'Osha' and 'Yerba Mansa' of New Mexico," *AJP* 39 (1867): 202–6.

Lacour-Gayet, Robert. *Everyday Life in the United States before the Civil War, 1830–1860.* Translated by Mary Ilford. New York: Frederick Ungar, 1969.

Laidley, Joseph. "Practical Notes on Pharmacy," *AJP* 26 (1854): 100–106.

Lassaigne, M. "A New Process in Eudiometry for Calculating the Volume of the Elements of Atmospheric Air in Relation to Each Other," *AJP* 17 (1846): 294–96.

"Late Professor Proctor [*sic*], The," (Philadelphia) *Public Ledger,* February 13, 1874, p. 1.

Leamy, James C. "On Extractum Cimicufugae Fluidum," *AJP* 27 (1855): 515.

"Lectures in the Philadelphia College of Pharmacy," *AJP* 23 (1851): 196.

Lee, Charles O. "The First Courses in Pharmacy," *American Journal of Pharmaceutical Education* 4 (1940): 49–63.

Lehmann-Haupt, Hellmut. *The Book in America.* 2d ed. New York: Bowker, 1951.

Liebenau, Jonathan Michael. "Medical Science and Medical Industry, 1890–1929: A Study of Pharmaceutical Manufacturing in Philadelphia." Ph.D. diss., University of Pennsylvania, 1981.

"*Lilly Digest* . . . Operating Statements of 1947," *Journal of the American Pharmaceutical Association* (Prac. Pharm. Ed.) 9 (1948): 679–83.

"List of the Contributors to the Building Fund for the New Hall of the Philadelphia College of Pharmacy," *AJP* 40 (1868): 376–78.

Long, James W. "A Defence of Elixirs, Etc.," *AJP* 45 (1873): 53–57.

McCluggage, Robert W. *A History of the American Dental Association.* Chicago, 1959.

McElroy's Philadelphia Directory for 1844. Philadelphia: Biddle, 1844.

McElroy's Philadelphia Directory for 1845. Philadelphia: Biddle, 1845.

McElroy's Philadelphia Directory for 1846. Philadelphia: Biddle, 1846.

Macfarlan, J. F. "On the Use of Alcohol Mixed with Pyroxylic Spirit for Certain Pharmaceutical Preparations," *AJP* 28 (1856): 150–55.

McIntyre, William. "Pharmaceutical Colleges and Associations: Alumni Association of the Philadelphia College of Pharmacy," *AJP* 46 (1874): 140.

――――. "Tribute to the Memory of Professor Procter," *AJP* 76 (1905): 42–43.

Maclagan, Douglass. "The Essential Oil of Bitter Almonds," *AJP* 26 (1854): 344–48.

Mahla, F. "On Hydrastia," *AJP* 35 (1863): 433–36.

――――. "Adulteration of Medicines," *AJP* 40 (1868): 547–49.

Maisch, John M. "On the Adulteration of Drugs and Chemical Preparations," *AJP* 26 (1854): 210–11.

――――. "Physicians and Pharmaceutists, and Their Relations," *AJP* 28 (1856): 25–28, 110–13, 202–4.

――――. "On the Use of Our Indigenous Plants," *Proc APhA* 6 (1857): 153–61.

――――. "On the Proper Menstruum for Fluid Extracts," *AJP* 31 (1859): 305–8.

――――. "Notes on the Fluid Extracts of Buchu, Cimicifuga, Serpentaria, and Valerian," *AJP* 31 (1859): 312–17.

――――. "Some Facts in Relation to the Solubility of Phosphate of Iron," *AJP* 31 (1859): 410–14.

――――. "On Alumen Exsiccatum,"*AJP* 32 (1860): 16–21.

――――. "Gleanings from the German Journals," *AJP* 32 (1860): 42–47.

――――. "Examination of Oil of Peppermint," *AJP* 32 (1860): 105–9.

――――. "Analysis of Commercial Glacial Phosphoric Acid," *AJP* 32 (1860): 193–97.

――――. "Chemical Notes," *AJP* 32 (1860): 521–24.

――――. "On Chelidonium Majus, Lin.," *AJP* 33 (1861): 7–11.

――――. "On an Adulteration of Carmine," *AJP* 33 (1861): 17–18.

――――. "On the Chemical Constituents of Coca Leaves," *AJP* 33 (1861): 496–500.

――――. "On Commercial Belladonna Leaves," *AJP* 34 (1862): 123–29.

――――. "Report on the Progress of Pharmacy," *Proc APhA* 10 (1862): 49–191.

――――. "Impurities and Adulterations," *AJP* 36 (1864): 100–103.

――――. "On Colchicia," *AJP* 39 (1867): 97–104.

――――. "Carelessness in the Collection of Drugs," *AJP* 39 (1867): 304–06.

――――. "Editorial Department," *AJP* 39 (1867): 375–84, 472–80; *AJP* 43 (1871): 233–36, 330–34, 378–82, 426–31, 475–78, 563–64; *AJP* 44 (1872): 42–43, 136–40, 189–91, 233–36, 474–77; *AJP* 45 (1873): 139–42, 476–80.

――――. "On a Permanent Solution of Pyrophosphate of Soda and Iron," *AJP* 39 (1867): 388–95.

————. "Gleanings from German Journals," *AJP* 41 (1869): 417–22.
————. "On the Nomenclature and some Definitions in the Materia Medica List U.S.P.," *AJP* 41 (1869): 519–25.
————. "Pharmaceutical Legislation," *AJP* 42 (1870): 303–8.
————. "Ferrated Elixir of Cinchona," *AJP* 43 (1871): 219–21.
————, ed. "Pharmaceutical Colleges and Associations, *AJP* 43 (1871): 278–81.
[Maisch, John M.?]. "On Cremor Tartari Solubilis," *Drugg Circ* 2 (1858): 217.
Marks, James N. "First Annual Report of the Pharmaceutical Examining Board of Philadelphia," *AJP* 45 (1873): 84–85.
"Maryland College of Pharmacy," *AJP* 43 (1871): 177–78.
Mayer, F. F. "Gleanings from Foreign Journals," *AJP* 37 (1865): 27–32.
Mead, Charles, and A. P. Sharp. "Remarks on the United States Standard of Specific Gravity for Indicating the Strength of Alcohol, and on the Official Hydrometer," *AJP* 28 (1856): 209–12.
Meakim, John. "[Presidential Address]," *Proc APhA* 5 (1856): 5–7.
Medicus. "Criticism on the Plural of Formula," *AJP* 32 (1860): 131–32.
————. "Elegant Pharmacy," *AJP* 44 (1872): 161–62.
"Meeting of the American Pharmaceutical Association," *Drugg Circ* 1 (1857): 120–23.
Mercein, James R. "The Aesthetics of Labels," *AJP* 43 (1871): 255–57.
[Merrill, William S.]. "Remarks on the Active Principles of Hydrastis Canadensis," *AJP* 34 (1862): 308–10.
————. "On the Alkaloids of Hydrastis Canadensis," *AJP* 34 (1862): 495–503.
Meyer, Minnie Marie. "A History of Pharmaceutical Journals in the United States." B.Sc. thesis, University of Wisconsin, 1932.
Mill, James W. "Practical Knowledge," *AJP* 32 (1860): 12–16.
————. "Tinctura Ferri Chloridi," *AJP* 41 (1869): 457–61.
————. "The Strength of Fluid Extracts," *AJP* 43 (1871): 17–24.
Milleman, Philip. "On Cunila Mariana (American Dittany)," *AJP* 38 (1866): 495–96.
Miller, Adolph. "Tribute to the Memory of Professor Procter," *AJP* 77 (1905): 37–38.
Miller, Douglas T. *The Birth of Modern America, 1820–1850.* Indianapolis: Bobbs-Merrill, 1970.
Millington, P. "Notes on Funnels and Weights," *AJP* 36 (1864): 3–4.
"[Minutes of the Annual Meeting]," *Proc APhA* 6 (1857): 13.
"Minutes of the Committee on Historical Pharmacy," *Proc APhA* 52 (1904): 427–50.
"Minutes of the Convention of Delegates from Colleges and Societies of Pharmacy, Held in Baltimore on the 14th and 15th of September, Relative to Pharmaceutical Education," *AJP* 42 (1870): 500–504.
Minutes of the Convention of Pharmaceutists and Druggists Held in the City of New York, October 15, 1851. Philadelphia: Merrihew and Son, 1865.
"Minutes of the Eighteenth Annual Meeting," *Proc APhA* 18 (1870); 17–127.
"Minutes of the Eighth Annual Meeting," *Proc APhA* 8 (1859): 1–48.
"Minutes of the Eleventh Annual Meeting," *Proc APhA* 11 (1863): 15–59.
"Minutes of the Fifteenth Annual Meeting," *Proc APhA* 15 (1867): 17–122.
"[Minutes of the Fifth Annual Meeting]," *Proc APhA* 5 (1856): 3–29.

"[Minutes of the First Annual Meeting]," *Proc APhA* 1 (1852): 5–22.

"Minutes of the Fourteenth Annual Meeting," *Proc APhA* 14 (1866): 17–80.

"[Minutes of the Fourth Annual Meeting]," *Proc APhA* 4 (1855): 3–14.

"Minutes of the Maryland College of Pharmacy," *AJP* 29 (1857): 473–76.

"Minutes of the Nineteenth Annual Meeting," *Proc APhA* 19 (1871): 25–128.

"Minutes of the Ninth Annual Meeting," *Proc APhA* 9 (1860): 1–64.

"Minutes of the Pharmaceutical Meetings," *AJP* 15 (1843): 69–75, 234–36, 312–15; *AJP* 16 (1844): 70–74, 309–14; *AJP* 30 (1858): 573–74; *AJP* 43 (1871): 89–91, 134–38, 231–33; *AJP* 44 (1872): 88–89, 133–36, 278–79, 521–26; *AJP* 45 (1873): 42–44.

"Minutes of the Philadelphia College of Pharmacy," *AJP* 7 (1836): 343–44; *AJP* 8 (1836): 165–68, 254–56; *AJP* 9 (1837): 171–72; *AJP* 12 (1840): 74–86; *AJP* 13 (1841): 79–80; *AJP* 14 (1842): 167–75; *AJP* 15 (1843): 346–47; *AJP* 17 (1845): 73–74; *AJP* 18 (1846): 144–56; *AJP* 19 (1847): 154–60, 309–14; *AJP* 20 (1848): 142–47; *AJP* 21 (1849): 377–81; *AJP* 22 (1850): 90–93; *AJP* 23 (1851): 79–80, 282–85; *AJP* 25 (1853): 278–81; *AJP* 26 (1854): 283–85, 565–66; *AJP* 27 (1855): 279–82; *AJP* 28 (1856): 275–77, 568–69; *AJP* 29 (1857): 274–76, 569–71; *AJP* 30 (1858): 278–80, 571–73; *AJP* 31 (1859): 280–83, 581–83; *AJP* 32 (1860): 277–82, 565–70; *AJP* 33 (1861): 281–87, 572–75; *AJP* 36 (1864): 180–81, 273–83; *AJP* 37 (1865): 230–33; *AJP* 38 (1866): 276–78; *AJP* 39 (1867): 276–80, 561–65; *AJP* 40 (1868): 85–87, 277–81, 565–67; *AJP* 41 (1869): 273–80, 476; *AJP* 42 (1870): 275–79, 375–76, 564–68; *AJP* 43 (1871): 40–41, 179–83, 327–28, 473–75; *AJP* 44 (1872): 185afo3–88, 328–30, 467–69; *AJP* 45 (1873): 224–32, 377–78, 506–20; *AJP* 46 (1874): 132–33.

"[Minutes of the Second Annual Meeting]," *Proc APhA* 2 (1853): 3–23.

"Minutes of the Seventeenth Annual Meeting," *Proc APhA* 17 (1869): 17–122.

"Minutes of the Seventh Annual Meeting," *Proc APhA* 7 (1858): 13–48.

"Minutes of the Sixteenth Annual Meeting," *Proc APhA* 16 (1868): 17–128.

"Minutes of the Sixth Annual Meeting," *Proc APhA* 6 (1857): 5–34.

"Minutes of the Tenth Annual Meeting," *Proc APhA* 10 (1862): 9–48.

"[Minutes of the Third Annual Meeting]," *Proc APhA* 3 (1854): 3–13.

"Minutes of the Thirteenth Annual Meeting," *Proc APhA* 13 (1865): 17–96.

"Minutes of the Twelfth Annual Meeting," *Proc APhA* 12 (1864): 17–55.

"Minutes of the Twentieth Annual Meeting," *Proc APhA* 20 (1872): 25–113.

"Minutes of the Twenty-first Annual Meeting," *Proc APhA* 21 (1873): 25–113.

"Minutes of the Twenty-second Annual Meeting," *Proc APhA* 22 (1874): 463–575.

Mohr, Francis [Karl Friedrich], Theophilus Redwood, and William Procter, Jr. *Practical Pharmacy: The Arrangements, Apparatus, and Manipulations, of the Pharmaceutical Shop and Laboratory.* American ed. Philadelphia: Lea and Blanchard, 1849.

Mohr, Karl F. *Lehrbuch der pharmaceutischen Technik.* Braunschweig: Friedrich Bieweg, 1847.

Moore, J. B. "The Night-Bell," *AJP* 45 (1873): 339–46.

Moore, J. Faris. "Notice: American Pharmaceutical Association," *AJP* 36 (1864): 461.

Morfit, Campbell. *Chemical and Pharmaceutic Manipulations* . . . Philadelphia: Lindsay and Blakiston, 1849.

Mott, Frank Luther. *A History of American Magazines, 1741–1850.* Cambridge, Mass.: Harvard Univ. Press, 1930; reprint 1957.

———. *A History of American Magazines, 1850–1865.* Cambridge, Mass.: Harvard Univ. Press, 1938.

———. *A History of American Magazines, 1865–1885.* Cambridge, Mass.: Harvard Univ. Press, 1938.

Muller, Mr. "Preparation of Carbolic (Phenic) Acid," *AJP* 38 (1866): 18–19.

Munsel, W. "On the Preparation of Dammara Varnish," *AJP* 26 (1854): 317–19.

"National Medical Convention," *AJP* 11 (1840): 345–50.

[Newton, L. V.]. "College of Pharmacy," *Drugg Circ* 2 (1858): 197.

———. "Good News for the Rising Generation of Pharmaceutists," *Drugg Circ* 2 (1858): 217.

———. "Salutatory," *Drugg Circ* 3 (1859): 13.

———. "Aid and Exercise," *Drugg Circ* 3 (1859): 61.

———. "Accidents Will Happen," *Drugg Circ* 3 (1859): 109–10.

———. "The Pharmacopoeia and Its Revision," *Drugg Circ* 3 (1859): 133.

———. "The Business of the Apothecary," *Drugg Circ* 3 (1859): 205–6.

———. "Make Your Own Preparations," *Drugg Circ* 3 (1859): 277–78.

———. "What is a Drug Store Proper?" *Drugg Circ* 3 (1859): 278.

———. "Advertising," *Drugg Circ* 4 (1860): 13.

———. "Proceedings of the Pharmaceutical Association," *Drugg Circ* 4 (1860): 38.

———. "Sunday Store-keeping," *Drugg Circ* 4 (1860): 129–30.

———. "Sunday Shopkeeping," *Drugg Circ* 4 (1860): 194.

———. "Eclectic Pharmacy," *Drugg Circ* 4 (1860): 194–95.

———. "Concerning the Art of Putting Things," *Drugg Circ* 5 (1861): 37–38.

———. "References and Recommendation," *Drugg Circ* 5 (1861): 61.

———. "Dull Times," *Drugg Circ* 5 (1861): 85.

———. "New Year's Talk," *Drugg Circ* 6 (1862): 9.

———. "The Use of Calomel in the Army," *Drugg Circ* 7 (1863): 109.

———. "Kings County Physicians against the Apothecaries," *Drugg Circ* 7 (1863): 141.

———. "Proceedings of the American Pharmaceutical Association," *Drugg Circ* 8 (1864): 48.

———. "Death of Dr. Franklin Bache," *Drugg Circ* 8 (1864): 67.

———. "Indigenous Drugs," *Drugg Circ* 8 (1864): 67.

———. "A Word about the 'Circular,'" *Drugg Circ* 8 (1864): 87.

———. "Looking through a Brick," *Drugg Circ* 8 (1864): 87–88.

———. "Chemical versus Pharmaceutical Manufactures," *Drugg Circ* 9 (1865): 210.

"New York College of Pharmacy, The," *Drugg Circ* 8 (1864): 28.

"Non-Officinal Formulae in Local Use," *AJP* 40 (1868): 237–39.

Norwood, William F. *Medical Education in the United States before the Civil War.* Philadelphia: Univ. of Pennsylvania Press, 1944.

"Notes and News," *AJP* 78 (1906): 352, 401–2, 452.

"Notes on Pharmaceutical Education," *Proc APhA* 40 (1892): 309–22.

"Notice: To the Pharmaceutists and Druggists of the United States," *AJP* 27 (1855): 283.

Nye, Russel B. *Society and Culture in America, 1830–1860.* New York: Harper and Row, 1974.

"Obituary: Prof. Wm. Proctor [*sic*], Jr.," *The Pharmacist* 7 (1874): 93–94.

"Obituary, Dr. Theophilus Redwood," *AJP* 64 (1892): 223–24.

Oldberg, Oscar. "The Drug Business in Sweden," *AJP* 42 (1870): 22–29.

————. "Calomel," *AJP* 43 (1871): 154–56.

————. *A Course of Home Study for Pharmacists.* Chicago: Apothecaries' Company, 1891.

"On Chemical Substitutes for the Fermentation of Bread," *AJP* 26 (1854): 42–46.

"On Some of the Applications of Glycerine," *AJP* 33 (1861): 158–62.

"On the Chemical Composition of Magenta," *AJP* 35 (1863): 342–47.

"On the Composition of Vegetable Gums," *AJP* 32 (1860): 366–69.

"On the Manufacture of Ammonia and Ammoniacal Salts," *AJP* 26 (1854): 33–39.

"On the Mode of Preparation of the Olea Cocta," *AJP* 27 (1855): 67–68.

Osborne, George E. "David Stewart, M.D.: First American Professor of Pharmacy," *Am J Pharm Ed* 23 (1959): 219–30.

Parrish, Edward. "Eclectic Pharmacy," *AJP* 23 (1851): 329–35.

————. "The Pharmaceutical Convention of 1852," *AJP* 24 (1852): 133–36.

————. "On Pills of Sulphate of Quinia," *AJP* 25 (1853): 291–93.

————. "American Pharmacy," *AJP* 26 (1854): 115–18, 211–17, 289–92.

————. *An Introduction to Practical Pharmacy: Designed as a Textbook for the Student, and as a Guide to the Physician and Pharmaceutist.* Philadelphia: Blanchard and Lea, 1855.

————. "Pharmacy as a Business," *Proc APhA* 5 (1856): 59–68.

————. "Ethical Analysis," *Proc APhA* 6 (1857): 146–50.

————. "A Critical Notice of the Proceedings of the Pharmaceutical Association," *Drugg Circ* 2 (1858): 232–33.

————. "What Shall the Boys Do, These Times," *Drugg Circ* 5 (1861): 155.

————. "Prefatory Notes," *Proc APhA* 10 (1862): 3–7.

————. "On the Use of Funnels in Displacement," *AJP* 31 (1859): 327–30.

————. "On the Publication of the Revised Pharmacopoeia," *AJP* 34 (1862): 512–14.

————. "A Plea for the Handmaiden," *Proc APhA* 11 (1863): 271–73.

————. "A Discourse on Titles, etc.," *AJP* 39 (1867): 239–46.

————. "On the Relations of the Several Classes of Druggists and Pharmacists to the Colleges of Pharmacy," *AJP* 43 (1871): 481–85, 532–35.

————. "Illustrations of Some Pharmaceutical Processes and Apparatus, as Exhibited to the Class in the Philadelphia College of Pharmacy," *AJP* 44 (1872): 1–6.

————. "An Unusual Case, with Comments," *AJP* 44 (1872): 385–87.

Parrish, Edward, and William C. Bakes. "Fancy and Fashion in Pharmacy," *AJP* 33 (1861): 1–6.

Parrish, Edward, Israel Grahame, and Charles T. Carney. "Report on the Revision of the Pharmacopoeia," *Proc APhA* 8 (1859): 217–45.

Parrish, Edward, William Procter, Jr., and Ambrose Smith. "Report of the Philadelphia College of Pharmacy . . . regarding the Pharmaceutical Statistics of Pennsylvania," *Proc APhA* 2 (1853): 26–33.

Parrish, Edward, A. M. Stevens, and C. B. Guthrie. "Note by the Executive Committee," *Proc APhA* 3 (1854): 2.

Pasteur, L[ouis]. "On the Origin of Ferments," *AJP* 33 (1861): 165–70.

———. "Researches on the Molecular Dissymmetry of Natural Organic Products," *AJP* 34 (1862): 1–16, 97–112.

———. "Researches on Acetic Fermentation," *AJP* 37 (1865): 343–44.

Peck, S. P. "Letter Relating to the Sale of Poisons in Vermont," *AJP* 25 (1853): 207–9.

Percy, S. R. "Note Relative to the Effects of Atropia," *AJP* 41 (1869): 203–4.

Perkins, [sic] W. H. "On Mauve, or Aniline Purple," *AJP* 36 (1864): 171–73.

Perrins, J. Dyson. "On Hydrastine, an Alkaloid Occurring in Hydrastis Canadensis," *AJP* 34 (1862): 360–62.

Pessen, Edward. *Jacksonian America: Society, Personality, and Politics*. Homewood, Ill.: Dorsey Press, 1978.

Phares, D. L. "Viburnum Prunifolium in the Treatment of Threatened Abortion," *AJP* 39 (1867): 259–60.

"Pharmaceutical Colleges and Associations," *AJP* 44 (1872): 39–42, 177–81, 469–74, 522–26; *AJP* 45 (1873): 87–91, 179–90, 378–82; *AJP* 46 (1874): 135, 140.

Pharmacopoeia of the Massachusetts Medical Society. Boston: E. and J. Larkin, 1808.

Pharmacopoeia of the United States of America. Philadelphia: Grigg and Elliot, 1842; Philadelphia: J. B. Lippincott, 1863.

"Philadelphia College of Pharmacy," *AJP* 43 (1871): 172–76.

"Philadelphia Correspondence," *Drugg Circ* 10 (1866): 81–82.

Plummer, John T. "Proximative Analysis of a Concretion of Hairs Found in the Esophagus of a Slaughtered Ox," *AJP* 25 (1853): 101–4.

Polanyi, Karl. *The Great Transformation*. Boston: Beacon Press, 1967.

Polk, C. G. "Elixirs," *AJP* 45 (1873): 6–8.

———. "Patent Medicines and Private Formulas," *AJP* 45 (1873): 57–59.

Polli, M. "On Fermentation as a Cause of Various Diseases," *AJP* 35 (1863): 523–26.

Porter, Glenn, and Harold C. Livesay. *Merchants and Manufacturers*. Baltimore: Johns Hopkins Univ. Press, 1971.

Postell, William D. *The Development of Medical Literature*. New Orleans: Louisiana State Univ. School of Medicine, 1943.

Power, Frederick B. *Address Delivered before the Alumni Association of the Philadelphia College of Pharmacy*. Philadelphia, 1882.

Praag, D. "Veratria and Aconitia considered in a Toxicological and Pharmaco-Dynamical Point of View," *AJP* 28 (1856): 117–18.

Prescott, A. B. "Simple Apparatus for Rapid Evaporization at Limited Heat, under Reduced Pressure, without the Use of a Pump," *AJP* 42 (1870): 349–52.

———. "Pharmaceutical Education," *Proc APhA* 19 (1871): 425–29.

Preston, David. "On Capsicum Annuum," *AJP* 37 (1865): 161–65.

"Proceedings of the American Pharmaceutical Convention," *AJP* 25 (1853): 481–508.

"Proceedings of the Convention of Pharmaceutists and Druggists, Held in the City of New York, Oct. 15, 1851," *AJP* 24 (1852): 22–27.

"Proceedings of the National Convention for Revising the Pharmacopoeia of the U. States," *AJP* 32 (1860): 370–75.

"Proceedings of the National Pharmaceutical Convention," *AJP* 25 (1853): 1–19.

"Proceedings of the New York College of Pharmacy," *AJP* 17 (1845): 148–52; *AJP* 18 (1846): 245–54.

"Proceedings of the Philadelphia College of Pharmacy, in Relation to the Inspection of Drugs," *AJP* 25 (1853): 297–304.

Procter, Wallace. "On the Fruit of Magnolia Tripetala," *AJP* 44 (1872): 145–51.

"Procter Memorial Unveiled," *Journal of the American Pharmaceutical Association* [Practical Pharmacy edition] 2 (1941): 167.

"Publishing Committee," *AJP* 13 (1841): i.

Quin, Charles W. "Drops," *AJP* 36 (1864): 522–27.

Rau, Robert. "An Essay on Senna and Its Active Principle," *AJP* 38 (1866): 193–200.

Redwood, Theophilus, and Francis [Karl Friedrich] Mohr. *Practical Pharmacy: The Arrangements, Apparatus, and Manipulations, of the Pharmaceutical Shop and Laboratory.* London: Taylor, Walton, and Maberly, 1849.

"Relation of the Chemical Constitution and Physicological Action of Medicine. Addition of Iodide of Methyl to Vegetable Alkaloids," *AJP* 40 (1868): 440–42.

Remington, Joseph P. *The Practice of Pharmacy.* 2d ed. Philadelphia: J. B. Lippincott, 1893.

———. "A Memorial of William Procter, Jr.," *Proc APhA* 48 (1900): 22–28.

———. "Galenical Pharmacy during the Century," *Drugg Circ* 36 (1900): 166–68.

———. "Edward Robinson Squibb, M.D.," *AJP* 73 (1901): 419–31.

———. "Tributes to the Memory of Professor Procter," *AJP* 77 (1905): 36–37.

———. "Procter Reminiscences," *Proc APhA* 54 (1906): 549–61.

———. "Statement of Prof. Joseph P. Remington of the Philadelphia College of Pharmacy, Philadelphia, Pa." In *Hearings before the Committee of the Library, H.R. 11076, February 18, 1916.* Washington, D.C.: GPO, 1916.

"Report of a Joint Committee of the Philadelphia County Medical Society and the Philadelphia College of Pharmacy, Relative to Physicians' Prescriptions," *AJP* 24 (1852): 27–32.

"Report of the Committee of Revision, on the New Pharmacopoeia, Made to the College, at a Special Meeting, Nov. 6th, 1841," *AJP* 13 (1842): 265–87.

"Report of the Committee on Adulterations and Sophistications of Drugs, Medicines, Chemicals, etc.," *AJP* 23 (1851): 18–21

"Resolutions, etc., Adopted at a Meeting of the Board of Trustees of the Massachusetts College of Pharmacy, Nov. 15, 1852," *AJP* 25 (1853): 22.

"Review. Chemical and Pharmaceutic Manipulations . . . by Campbell Morfit," *AJP* 21 (1849): 102–14.

Reynolds, H. P. "On Campbell's Method of Percolation," *AJP* 41 (1869): 525–27.

———. "On Mistura Cretae," *AJP* 42 (1870): 391–92.

Rice, Charles. "Ethereal Solution of Quinia," *AJP* 43 (1871): 303–7.

Richardson, W. D., Jr. "On Lobelina," *AJP* 44 (1872): 293–95.

Risse, Guenter B. *Medicine Without Doctors*. New York: Science History Publications, 1977.

Robertson, Andrew, T. Smith, and H. Smith. "On Extract of Indian Hemp," *AJP* 19 (1847): 195–98.

Rorabaugh, W. J. *The Craft Apprentice*. New York: Oxford Univ. Press, 1986.

Rose, Heinrich. "Preparation of Aluminium from Cryolite," *AJP* 28 (1856): 237–42.

Rosen, George. "Fees and Fee Bills: Some Economic Aspects of Medical Practice in Nineteenth Century America," *Bulletin of the History of Medicine,* supp. 6 (1946).

Rosenberg, Charles E. "Social Class and Medical Care in Nineteenth-Century America: The Rise and Fall of the Dispensary," *Journal of the History of Medicine* 29 (1974): 32–54.

———. *No Other Gods*. Baltimore: Johns Hopkins Univ. Press, 1976.

———, ed. *The Therapeutic Revolution: Essays in the Social History of American Medicine*. Philadelphia: Univ. of Pennsylvania Press, 1979.

Roth, Julius. "Professionalism: The Sociologist's Decoy," *Sociology of Work and Occupations* 1 (1974): 18–32.

Rothstein, William G. *American Physicians in the Nineteenth Century*. Baltimore: Johns Hopkins Univ. Press, 1977.

[Royal Society of London]. *Catalogue of Scientific Papers . . . .* New York: Johnson Reprint, 1965.

Rusby, H. H. "Fifty Years of Materia Medica," *Drugg Circ* 51 (1907): 29–43.

S. "Art. LIX.—Pharmaceutical Essays—No. 1," *AJP* 6 (1834): 279–82.

Sadtler, Samuel P. "Influence of Pharmacists on the Development and Advance of Modern Chemistry," *AJP* 93 (1921): 197–207.

Sale, Luther E. "Cucumber Ointment," *AJP* 43 (1871): 27.

Salmon, Lucy M. *The Newspaper and the Historian*. New York: Oxford Univ. Press, 1923.

Sargent, E. H. "On Fluid Extracts," *AJP* 42 (1870): 337–42.

Saum, Lewis. *The Popular Mood of Pre-Civil War America*. Westport, Conn.: Greenwood Press, 1980.

Saxe, John G. *The Poetical Works of John Godfrey Saxe*. Boston: Houghton Mifflin, 1892.

Schumann, Th. "The Relation between the Physician and Pharmaceutist," *AJP* 44 (1872): 401–4.

Scott, Donald M. "The Professional that Vanished: Public Lecturing in Mid-Nineteenth-Century America." In *Professions and Professional Ideologies in America,* edited by Gerald L. Geison. Chapel Hill: Univ. of North Carolina Press, 1983.

Scott, John Mark. "K. F. Mohr, Father of Volumetric Analysis," *Chymia* 3 (1950): 191–203.

Scoville, Wilbur. "A Procter Memorial," *AJP* 73 (1901): 86–88.

"Second Session—Tuesday Morning, May 8, 1900," *Proc APhA* 48 (1900): 20–58.

Sedlak, Michael W. *American Educational History: A Guide to Information Sources*. Detroit: Gale, 1981.

"Semi-Centennial Anniversary of the Philadelphia College of Pharmacy," *AJP* 43 (1871): 130–34.

"Seventy-first Annual Meeting of the American Pharmaceutical Association . . . Abstract of Minutes," *Journal of the American Pharmaceutical Association* 12 (1923): 893–911.

Shafer, Henry B. *The American Medical Profession, 1783–1850.* New York: Columbia Univ. Press, 1936.

Shryock, Richard Harrison. *Medicine and Society in America: 1600–1860.* Ithaca, N.Y.: Great Seal Books (Cornell), 1960.

Shinn, James T. "Tribute to the Memory of Professor Procter," *AJP* 77 (1905): 40–41.

S[mith?], A[mbrose]. "An Introduction to Practical Pharmacy . . . by Edward Parrish," *AJP* 28 (1856): 10–20.

Smith, Daniel B. "Address Delivered to the Graduates of the Philadelphia College of Pharmacy, May 8th, 1837," *AJP* 9 (1837): 89–98.

——. "Chemical and Pharmaceutical Manipulations, 2nd edition, . . . by Campbell Morfit," *AJP* 29 (1857): 17–27.

Smith, T., and H. Smith. "On an Antidote at Once for Prussic Acid, Antimony, and Arsenic," *AJP* 38 (1866): 12–18.

Smith, Wm. Manlius. "At Attempt to Answer the Question: Which Part of the Plant Conium Maculatum Is the Best for Medicinal Use," *AJP* 40 (1868): 459–66.

Society of Friends in Pennsylvania. *Inventory of Church Archives.* Philadelphia, Friends Historical Association, 1941.

"Soluble Citrate of Magnesia," *AJP* 31 (1859): 407–8.

Sonnedecker, Glenn. "American Pharmaceutical Education before 1900." Ph.D. diss., University of Wisconsin, 1952.

——. "The Conference of Schools of Pharmacy," *Am J Pharm Ed* 18 (1954): 389–401.

——. "The Rise of Drug Manufacture in America," *Emory University Quarterly* 21 (1965): 73–87.

——, revisor. *Kremers and Urdang's History of Pharmacy.* 4th ed. Philadelphia: J. B. Lippincott, 1976.

"Special Notices," (Philadelphia) *Public Ledger,* February 19, 1874.

"Sponge Divers of Calymnos, The," *AJP* 38 (1866): 60–61.

Sprowls, Joseph B., Jr., and Harold M. Beal, eds. *American Pharmacy: An Introduction to Pharmaceutical Technics and Dosage Forms.* 6th ed. Philadelphia: J. B. Lippincott, 1966.

Squibb, Edward R. "Preparation of Citrate of Iron and Quinia, and Its Constituents," *AJP* 27 (1855): 294–300.

——. "Spiritus Aetheris Nitrici," *AJP* 28 (1856): 289–304.

——. "The Manufacture, Impurities, and Tests of Chloroform," *AJP* 29 (1857): 430–41.

——. "The Process of Percolation," *AJP* 30 (1858): 97–102.

——. "Weights and Measures of the Pharmacopoeia," *AJP* 32 (1860): 26–29.

——. "Observations upon Some Formulae and Processes for Preparations that May Be Brought Forward for Admission into the Pharmacopoeia," *AJP* 32 (1860): 29–42.

————. "Extractum Cinchonae Fluidum (Containing Aromatics)," *AJP* 35 (1863): 230–32.

————. "Proposed Economy of Alcohol in Percolation, as Applied to the Extracts and Fluid Extracts of the Pharmacopoeia," *AJP* 38 (1866): 109–28.

————. "Advice upon Epidemic Cholera," *AJP* 38 (1866): 310–14.

————. "Improved Process for Officinal Fluid Extract of Buchu," *AJP* 39 (1867): 129–35.

————. "Pharmacy of the Cinchonas," *AJP* 39 (1867): 289–303, 398–414, 513–29.

————. "Liquor Opii Compositus (Compound Solution of Opium)," *AJP* 42 (1870): 33–60.

Stabler, R. H. "Practical Observations," *AJP* 23 (1851): 121–24.

————. "On Podophyllin," *AJP* 30 (1858): 508–12.

Starr, Paul. *The Social Transformation of American Medicine*. New York: Basic Books, 1982.

Stearns, Frederick. "On Compound Syrup of Ipecacuanha and Seneka," *AJP* 28 (1856): 205–6.

————. "Upon Improvements in Methods of Rendering Medicinal Preparations Pleasing to the Eye and to the Taste, and Agreeable to Use," *Proc APhA* 6 (1857): 134–45.

Stenhouse, John. "On Charcoal as a Disinfectant," *AJP* 27 (1855): 167–72.

Stieren, Edward. "Remarks on the Iodine-Water of Dr. Anders," *AJP* 29 (1857): 394–97.

Stuart [Stewart], David. "On Nascent Manures," *AJP* 27 (1855): 246–51.

Sweringen, Hiram V. "Our Next Pharmacopoeia," *AJP* 42 (1870): 110–13.

Taylor, Alfred B. "Weights of the Pharmacopoeia," *AJP* 32 (1860): 97–104.

————. "Review of the United States Pharmacopoeia," *AJP* 35 (1863): 401–20.

Taylor, Horace B. "On Capsicum Annuum (An Inaugural Essay)," *AJP* 29 (1857): 303–6.

Taylor, Samuel. "Hints to Dispensers," *AJP* 38 (1866): 347–52.

Thayer, Henry. "On the Preservation of Fluid Extracts," *AJP* 31 (1859): 216–22.

Thomas, N. Spencer. "A Comparison of Dr. Squibb's Plan for Economizing Alcohol and N. Spencer Thomas's Patented Process for Making Fluid Extracts," *AJP* 38 (1866): 218–24.

Thrush, M. Clayton. "The United States Pharmacopoeia and Its Predecessors," *Drugg Circ* 51 (1907): 43–52.

Tomlinson, Charles. "Why Do Bees Work in the Dark," *Drugg Circ* 10 (1866): 161–62.

"To the Members and Graduates of the Philadelphia College of Pharmacy," *AJP* 12 (1840): 86–87.

Trease, George E., and Wm Charles Evans. *Pharmacognosy*. 11th ed. London: Bailliere Tindell, 1978.

Turner, Wm. L. "Note on Lozenge Cutting," *AJP* 39 (1867): 206.

————. "The Doctor and the Apothecary," *AJP* 43 (1871): 149–51.

————. "The Need of Practical Information in Our Pharmaceutical Publications," *AJP* 44 (1872): 531–35.

Unveiling Ceremonies of the Painting "The Father of American Pharmacy" by Dean Cornwell, N.A. Philadelphia: n.p. [Wyeth?], 1943.

Urdang, George. "The First Century of the Pharmaceutical Society of Great Britain," *Journal of the American Pharmaceutical Association* 3 (1942): 420–27.

———. "The Reagent Bottles of William Procter, Jr.," *Pharmaceutical Archives* 14 (1943): 45–48.

———. "The Development of the Pharmaceutical Text," *Am J Pharm Ed* 8 (1944): 328–33.

———. "College of Pharmacey Associations," *Am J Pharm Ed* 8 (1944): 333–39.

———. "The Influence of the Quakers on Philadelphia Institutions," *AJP* 118 (1846): 81–89.

———. "The Rescue of the U.S. Pharmacopoeia by Organized American Pharmacy in the Eighteen Seventies," *Am J Pharm Ed* 15 (1950): 172–84.

———. "Edward Parrish, A Forgotten Pharmaceutical Reformer," *Am J Pharm Ed* 14 (1950): 223–32.

———. "The Part of Doctors of Medicine in Pharmaceutical Education," *Am J Pharm Ed* 14 (1950): 546–56.

Wadgymar, A. "Deportment of the Most Important Alkaloids with Reagents, and a Systematic Method of Effecting the Detection of these Substances," *AJP* 38 (1866): 447–54.

Wallace, J. M. "Antidote to Corrosive Sublimate," *AJP* 15 (1843): 332–33.

Wankmueller, Armin. "Deutschsprachige Lehrbuecher der pharmazeutischen Technologie von 1845 bis 1945," *Deutsche Apotheker Zeitung* 122 (1982): 1236–39.

Warner, John Harley. *The Therapeutic Perspective: Medical Practice, Knowledge, and Identity in America, 1820–1885.* Cambridge, Mass.: Harvard Univ. Press, 1986.

Weinstein, M. M., M. B. Mrtek, R. L. Lambert, and R. G. Mrtek, ". . . From These Ashes," *Pharmacy in History* 15 (1973): 54–65, 107–16.

Whorton, James. "Chemistry." In *The Education of American Physicians: Historical Essays,* edited by Ronald L. Numbers. Berkeley: Univ. of California Press, 1980.

Wiegand, Thomas S. "Tribute to the Memory of Professor Procter," *AJP* 77 (1905): 43–44.

Wiegand, Thomas S., and John M. Maisch. "Prefatory Notice," *Proc APhA* 21 (1873): 21–22.

"William Procter, Jun.," *Pharmaceutical Journal and Transactions* 33 (1874): 707.

Wimmer, Curt. *The College of Pharmacy at the City of New York.* New York, 1929.

Windholz, Martha, ed. *The Merck Index.* Rahway, N.J.: Merck, 1983.

Woehler, Prof. "On the Reduction of Iron by Hydrogen," *AJP* 28 (1856): 139–40.

Wood, C. H. "Pharaoh's Serpents," *AJP* 38 (1866): 61–62.

———. "On Hydrargyri Iodidum Viride, B. P.," *AJP* 40 (1868): 337–42.

Wood, George B. "Lecture Introductory to the Course on Materia Medica and Pharmacy in the University of Pennsylvania. Session, 1835–6," *AJP* 8 (1836): 286–300.

Wood, George B., and Franklin Bache. *The Dispensatory of the United States of America.* Philadelphia: J. B. Lippincott, 1865; 13th ed., Philadelphia: J. B. Lippincott, 1870.

Wood, H. C., Joseph P. Remington, and Samuel P. Sadtler. *Dispensatory of the United States of America*. 17th ed. Philadelphia: J. B. Lippincott, 1894.

Wormley, Theo. G. "A Contribution to Our Knowledge of the Chemical Composition of Gelsemium Sempervirens," *AJP* 42 (1870): 1–16.

Worthington, Henry W. "On Veratrum Viride, with Experimental Facts, to Serve as a Chemical History of the Root of this Plant," *AJP* 10 (1838): 89–97.

Wright, Saml. P. "On Suppositories," *AJP* 42 (1870): 197–200.

Wruble, Milton. *Studies in Percolation*. Reprint. Philadelphia: American Journal of Pharmacy, 1933. (Originally published in *AJP* 105 [1933].)

Young, James Harvey. *Pure Food: Securing the Federal Food and Drugs Act of 1906*. Princeton, N.J.: Princeton Univ. Press, 1989.

Zochert, Donald. "Science and the Common Man in Ante-bellum America," *Isis* 65 (1974): 448–93.

Index

About the Author

Gregory J. Higby is Director of the American Institute of the
History of Pharmacy and Adjunct Assistant Professor, University
of Wisconsin-Madison School of Pharmacy. He received a bach-
elor's degree from the University of Michigan and his master's
degree and doctorate from the University of Wisconsin.